Medical Tourism

Medical Tourism

Editor

Regina Belo

scitus
academics

Medical Tourism
Edited by **Regina Belo**

ISBN: 978-1-68117-550-8
Library of Congress Control Number: 2016942956

www.scitusacademics.com
Box No. 4766, 616 Corporate Way,
Suite 2, Valley Cottage,
NY 10989

Notice

Reasonable efforts have been made to publish reliable data and views articulated in the chapters are those of the individual contributors, and not necessarily those of the editors or publishers. Editors or publishers are not responsible for the accuracy of the information in the published chapters or consequences of their use. The publisher believes no responsibility for any damage or grievance to the persons or property arising out of the use of any materials, instructions, methods or thoughts in the book. The editors and the publisher have attempted to trace the copyright holders of all material reproduced in this publication and apologize to copyright holders if permission has not been obtained. If any copyright holder has not been acknowledged, please write to us so we may rectify.

Printed in United States of America on Acid Free Paper

Preface

Medical tourism represents a worldwide, multibillion-dollar phenomenon that is expected to grow considerably in the next decade. For the individual interested in health services, cost is the key factor involved in the decision to receive medical care abroad. Today we are experiencing both qualitative and quantitative shifts in patient mobility, as people travel from richer to less-developed countries in order to access health services. Such shift is mostly driven by the relative low-cost of treatments in less developed nations, the availability of inexpensive flights and increased marketing and online consumer information about the availability of medical services. What really puts the word "tourism" in medical tourism concept is that people often stay in the foreign country after the medical procedure. Travelers can thus take advantage of their visit by sightseeing, taking day trips or participating in any other traditional tourism activities. Categories of different treatments and their availability also represent an important factor in decision to engage in medical tourism. The most common types of procedures that patients pursue during medical tourism trips are elective cosmetic surgery, dentistry, organ transplantation, cardiac surgery and orthopedic surgery. However, a wide variety of services can be obtained through medical tourism, ranging from various essential treatments to different kinds of traditional and alternative treatments.

Drawing on a range of theoretical and methodological perspectives, this book titled *Medical Tourism* critically address the substantial political, philosophical and ethical issues that arise out of the transnational practices of medical tourism.

Table of Contents

Chapter 1

What is known about the patient's experience of medical tourism? A scoping review

Valorie A Crooks[1*], Paul Kingsbury[1], Jeremy Snyder[2] and Rory Johnston[1]

[1]Department of Geography, Simon Fraser University, 8888 University Drive, Burnaby, British Columbia, Canada.

[2]Faculty of Health Sciences, Simon Fraser University, 8888 University Drive, Burnaby, British Columbia, Canada.

ABSTRACT

Background

Medical tourism is understood as travel abroad with the intention of obtaining non-emergency medical services. This practice is the subject of increasing interest, but little is known about its scope.

Methods

A comprehensive scoping review of published academic articles, media sources, and grey literature reports was performed to answer the question: what is known about the patient's experience of medical tourism? The review was accomplished in three steps: (1) identifying the question and relevant literature; (2) selecting the literature; (3) charting, collating, and summarizing the information. Overall themes were identified from this process.

Results

291 sources were identified for review from the databases searched, the majority of which were media pieces (n = 176). A further 57 sources were included for review after hand searching reference lists. Of the 348 sources that were gathered, 216 were ultimately included in this scoping review. Only a small minority of sources reported on empirical studies that involved the collection of primary data (n = 5). The four themes identified via the review were: (1) decision-making (e.g., push and pull factors that operate to shape patients' decisions); (2) motivations (e.g., procedure-, cost-, and travel-based factors motivating patients to seek care abroad); (3) risks (e.g., health and travel risks); and (4) first-hand accounts (e.g., patients' experiential accounts of having gone abroad for medical care). These themes represent the most discussed issues about the patient's experience of medical tourism in the English-language academic, media, and grey literatures.

Conclusions

This review demonstrates the need for additional research on numerous issues, including: (1) understanding how multiple information sources are consulted and evaluated by patients before deciding upon medical tourism; (2) examining how patients understand the risks of care abroad; (3) gathering patients' prospective and retrospective accounts; and (4) the push and pull factors, as well as the motives of patients to participate in medical tourism. The findings from this scoping review and the knowledge gaps it uncovered also demonstrate that there is great potential for new contributions to our understanding of the patient's experience of medical tourism.

1. BACKGROUND

Medical tourism is becoming an increasingly popular option for patients looking to access procedures (typically via out-of-pocket payment) that are seemingly unavailable to them in their home countries due to lack of affordability, lack of availability, and/or lengthy waiting lists, among other reasons [1, 2]. In its broadest conceptualization, medical tourism refers to "travel with the express purpose of obtaining health services abroad" (p.193) [3]. People wishing to access procedures such as cardiac, orthopaedic, dental, and plastic surgeries are going to key destination countries known to provide care for international patients [4]. For example, countries such as India, Singapore, and Thailand have become global leaders in the industry, providing services for patients from around the world. Brochures, websites, and other marketing materials promote the services of hospitals in these countries wanting to attract international patients [5]. Facilitators/brokers specializing in medical tourism further promote the practice, and offer services such as making travel bookings, assisting with selecting hospitals and surgeons abroad, and helping with completing paperwork to potential medical tourists [6].

The practice of medical tourism does not exist without criticism, particularly when involving patients from developed nations going to developing nations for procedures. It is thought to contribute to the commodification of health and health care by allowing those with the financial means to do so to purchase care that may be unavailable to other citizens [7]. The practice can also lead to international patients receiving a higher standard of care than residents of the country where it is being given [8]. Another criticism is that health service providers trained in countries with publicly-funded education systems who are involved in privately treating international patients are misdirecting the public funds that contributed to their training [9]. It has, however, been suggested that if the industry is properly regulated, medical tourism can provide a viable means by which developing countries can gain access to needed revenue and developed countries can lessen 'bottlenecks' in their health systems [10]. The presence of medical tourism hospitals in developing nations is also thought to lessen the international brain drain of health human resources by providing surgeons and others with access to advanced, high technology work environments [3].

Although estimates of the number of patients engaging in medical tourism each year vary widely, ranging from millions to tens of thousands, there has been speculation that growth in the industry will continue in the coming years [11, 12, 13]. Given the prominence of this global industry, research and media attention focused on it will also surely continue to grow. With projections of growth in the industry and the existence of significant criticisms about the practice, it is an opportune time to undertake knowledge syntheses to assess what exactly is known about medical tourism so as to ultimately inform research, government, and industry agendas alike. In the remainder of this article we take on this task, presenting the findings of a scoping review that addresses the question: what is known about the patient's experience of medical tourism? This article serves as the first attempt to draw together what is known about this issue, and thus is a valuable contribution to the burgeoning literature on medical tourism. In an attempt to be as comprehensive and inclusive as possible, multiple types of sources are included in the review, including: academic articles, newspaper and magazine articles, industry reports, and law reviews. Such inclusivity is central to the scoping review process in general, where the aim is to appreciate the breadth of knowledge that is available on a particular topic [14].

While there is no singular definition of medical tourism that has gained wide acceptance, in this article we place some widely acknowledged parameters on what it is understood to be in order to focus the scoping review. People who become ill or injured while traveling abroad and require hospital care are not thought to be medical tourists, nor are expatriates accessing care in the countries or regions in which they live. A survey run by the Thai government to assess the scope of its domestic medical tourism industry distinguished between international patients who were medical tourists, ill vacationers, and expatriates living in Thailand or a neighbouring country, which confirms the distinctions being used here [15]. Established cross-border care arrangements between countries are not forms of medical tourism. This is because out-of-pocket payments for the accessed care are not typically made under such arrangements, as is the case for medical tourists, and because these arrangements typically require referrals to be given for care that is not available locally based on collaborative arrangements between hospitals or care systems. Meanwhile, medical tourists can choose to go abroad for care without the referral of a physician. These distinctions are made elsewhere. For example, a World Health Organization report on cross-border care within Europe distinguishes between patients travelling *independently* (i.e., without referral) for care internationally, those who are sent abroad by their home systems in order to access specialized care that is not available locally, and those who live in border regions with traditions of sharing care across borders [16]. Further, the pursuit of complementary and alternative care abroad is not medical tourism; instead, it falls under the even broader rubric of health tourism. When taken together, these parameters result in achieving a focused understanding of medical tourism, whereby it occurs when patients *intentionally* leave their country of residence *outside of established cross-border care arrangements* in pursuit of non-emergency medical interventions (namely surgeries) abroad that are commonly paid for out-of-pocket. This typically includes staying abroad for at least part of the recovery period, whereby such post-discharge time can be spent at tourist resorts that cater to international patients [17, 18, 19].

2. METHODS

Broadly speaking, knowledge syntheses aim to collect and evaluate the current state of knowledge on a particular issue [14]. The scoping review is a knowledge synthesis technique that is most commonly used when: it is difficult to identify a narrow review question; studies in the reviewed sources are likely to have employed a range of data collection and analysis techniques; no prior synthesis has been undertaken on the topic; and a quality assessment of reviewed sources is not going to be conducted [14]. In this article the findings of a scoping review that meets all of these criteria is presented. The review poses the broad question: what is known about the patient's experience of medical tourism? The synthesis presented in this article follows the scoping review protocol set out by Arksey and O'Malley [14]. In the remainder of this section we outline the steps undertaken to complete the review.

2.1 Identifying the Question and Relevant Literature

The first step was to develop the scoping question, which was done by holding a research team meeting to identify a potentially fruitful and also useful issue to focus on within the area of medical tourism. Next the team moved to delineate a search strategy that would lead to the identification of relevant literature. To do this, keywords were first identified based on review of relevant literature and ultimately team consensus. As depicted in Table 1, keywords probed five main categories: (1) focus; (2) what; (3) who; (4) why; and (5) where. Eight types of rationale were identified for the why category, as shown in Table 1. Known destination and departure countries were used to populate the where category.

Table 1. Scoping review keyword search strategy

Focus	What	Who	Why	Where
Medical tourism Health tourism	Surger* Elective surger* Surgical Procedure* Hospital* Clinic*	Patient Tourist	Decision making Factors Decision Attitudes Motivation Destination choice	Destination Brazil India Thailand South Africa Indonesia Cuba Mexico Philippines Singapore United States Canada
			Tour* Travel Vacation* Adventur*	
			Wait time Wait list Queue Speed	
			Value* Ethic*	
			Privat* Effects Two tier	
			Cost savings Affordability Savings Cost	
			Motivat* Perspective*	
			Distance Quality	

Following finalization of the keywords, a search strategy was created with the input of a librarian to scope the English-language academic, media, and grey literatures to achieve as much breadth as possible. Combinations of terms were rationally searched in 18 databases, summarized in Table 2, with a different search strategy being employed between academic and media databases. For academic databases, keywords across the five categories summarized in Table 1 were searched using Boolean operators in order to maximize the permutations of terms scoped. Certain combinations of keywords yielded unmanageably large and mostly irrelevant results. In these instances the search manager narrowed the results either by adding additional keywords or removing the term that had the broadest results to ultimately enhance the focus and relevance of the findings.

As searching all media sources around the globe was not feasible, the team focused the media search on a particular country (Canada) to capture local, regional, and national coverage of medical tourism. Team had best access to media search databases for Canadian sources, which is why Canada was selected as the focus. It is, however, expected that the identified sources may replicate what some media coverage of the patient's experience of medical tourism is like in other prominent patient departure countries (e.g., the United States [US], the United Kingdom), keeping in mind that there are important differences in regulatory and health system environments operating between such countries that are likely to have influences on patients' experiences. Only the terms 'medical tourism' and 'health tourism' were searched in the media databases. The databases were searched broadly for these terms, and key North American sources known to frequently cover Canadian health services issues were also specifically searched within the Lexis Nexis database (namely the New York Times, Time inc., Globe and Mail, Associated Press, magazines, Washington Post, Toronto Star, Toronto Sun, CBC News). Sources of all types across the academic and media databases deemed relevant to the search were retrieved and organized using the Refworks bibliographic management program.

2.2 Selecting the Literature

In order to select literature for inclusion in the review the team first searched titles and abstracts of the identified sources. All team members independently reviewed each title and abstract and consensus was sought as to whether or not to read sources in full, which were reviewed in batches given the large number of identified sources. '*Post hoc*' inclusion criteria were created and employed by the team at this step. The development of such criteria '*post hoc*' is central to the scoping review process as it is unlikely that researchers will be able to identify bases for exclusion at the outset, and this, in fact, is a key point of differentiation between the scoping and systematic review processes [14]. The team identified three bases for exclusion: (1) there was no focus on medical intervention, which included articles that dealt with health tourism more broadly such as international travel to healing spas; (2) there was an exclusive focus on 'reproductive tourism' or 'transplant tourism', as the medical intervention (if any) in such cases is not restricted to the international patient and thus raises separate considerations; and (3) there was an overly general focus on international trade in health services or cross-border care, where there seemed to be no explicit reference to medical tourism. Disagreements regarding whether or not a source should be included for full review were discussed among the team until consensus was reached. As the title and abstract review moved forward, the level of agreement among team members, which was already high from the outset, continued to increase.

Upon completion of the title and abstract review, included sources were reviewed in full. Titles and abstracts were not available for media sources and so they were not involved in the first round of inclusion/exclusion and were all reviewed in full. Further, included sources were hand searched and relevant sources not already gathered from the search databases were identified for full review. The three exclusion criteria created at the title and abstract review stage were applied at the full review stage, with one additional criterion being applied: if no 'informational points' (i.e., discrete pieces of information found within sources that contributed to answering the scoping question) were extracted from the source it was excluded. Two team members reviewed every source identified for full review. Sources were reviewed in batches, and upon completion of each new batch the team met to review decisions regarding the inclusion or exclusion of sources. As with the title and abstract review stage, any disagreement was resolved through seeking consensus among all members after discussion.

Table 2. Databases searched for scoping review

Database Type	Database	Temporal Period Covered
Academic	Academic Search Premier	1984 - 20/08/2009
	Ageline	1978 - 10/08/2009
	Biomed Central	no recorded start date - 10/08/2009
	Business Source Complete	no recorded start date - 20/08/2009
	Canada Research Index	1982 - 21/07/2009
	CINAHL	1982 - 20/07/2009
	CPI.Q	1988 - 22/10/2009
	EconLit	1969 - 9/08/2009
	Geobase	1980 - 9/08/2009
	Global Health	1973 - 10/08/2009
	Medline	1950 - 9/08/2009
	PAIS International	1972 - 9/08/2009
	PsycINFO	1887 - 10/08/2009
	Sociological Abstracts	1963 - 20/08/2009
	Web of Science	1900 - 10/08/2009
Media	Alternative Press Index	1991 - 20/07/2009
	CBCA Current Events	1982 - 21/07/2009
	Canadian Newstand	1985 - 22/10/2009
	Lexis Nexis	no recorded start date - 22/10/2009

2.3 Charting, Collating, and Summarizing the Information

To chart the informational points extracted from the sources, a spreadsheet was created and securely hosted online that was used by all team members. Details regarding publication information, study design (if relevant), and the sample (if relevant) were recorded, along with any informational points pertinent to the overall scoping question. These details were recorded independently by each reviewer for all of the sources, including those that were ultimately excluded. The extracted informational points were discussed during team meetings in order to work towards gaining an overall perspective on the themes emerging from the literature pertaining to the scoping question, which is essential to the charting process [14]. As discussed in the results section, four such themes were identified. Following this the extracted informational points in the spreadsheet were colour coded according to theme in order to assist with organizing the reporting of the scoping review findings. Finally, the team then worked together to identify important avenues for future research by identifying knowledge gaps.

3. RESULTS

Shown in Figure 1, 291 sources were identified for review from the databases searched, the majority of which were media pieces (n = 176). A further 57 sources were included for review after hand searching reference lists. Of the 348 sources that were reviewed either partially (title and abstract) or in full, 216 were ultimately included in the scoping review (a full list of included sources can be obtained by contacting the lead author). While this is a large number relative to some knowledge syntheses (e.g. many systematic reviews), as an aim of scoping reviews is to pose a broad question and then appreciate the extent of the literature on an issue from a variety of sources it is not unthinkable to include this many sources in a single review [14]. The included sources were authored in ten different countries (Australia, Canada, France, India, Malaysia, Norway, Spain, Thailand, US, and United Kingdom - many of which are prominent home or destination countries for medical tourists), and made explicit reference to a number of others (e.g., China, Cuba, Germany, Jordan, Mexico, Poland, Singapore, South Africa, Tunisia, United Arab Emirates, Yemen), which geographically reflects the state of interest in medical tourism.

Only a small minority of sources included in the scoping review were reporting on empirical studies that involved the collection of primary data (n = 5) [20, 21, 22, 23, 24]. Instead, they were mostly conceptual pieces, discussion papers, law reviews, commentaries, editorials, reviews, magazine and newspaper articles, reports, and business briefs published in a range of venues. The breadth of sources included in the review has allowed for perspectives from a number of sectors, including industry, health professionals, health service administrators, tourism operators, and academia, to be captured in the review. At the same time, most of the sources are heavily speculative or anecdotal in nature, relying on opinion instead of empirical facts. Because of the emerging nature of medical tourism, however, and the lag time involved in gathering and publishing empirical evidence, grey literature and media reports that offer preliminary glimpses were not excluded because they provide some of the only insights into the patient's perspective. Instead, caution was exercised in interpreting themes from the informational points extracted from the sources and a tentative tone is adopted below in sharing what was gleaned.

DETERMINATION OF SCOPING REVIEW QUESTION

↓

BROAD SEARCH OF LITERATURE TO DETERMINE KEYWORDS

↓

DETERMINATION OF KEYWORDS

↓

DEVELOPMENT OF SEARCH STRATEGY

↓

SEARCH 18 DATABASES USING KEYWORDS

↓

IDENTIFICATION OF ARTICLES FOR TITLE AND ABSTRACT REVIEW
115 articles and reports and 176 media pieces sourced from 18 databases
(minus duplicates)

↓

INDEPENDENT REVIEW OF TITLES AND ABSTRACTS

↓

TITLE AND ABSTRACT REVIEW MEETINGS
91 articles and reports and 176 media pieces selected for full review
(note: separate titles and abstracts were not available for media pieces
so they were all reviewed in full)

↓

INDEPENDENT ARTICLE REVIEW

↓

IDENTIFICATION OF ADDITIONAL ARTICLES THROUGH HAND SEARCHING
reference lists of reviewed articles and reports were hand searched for new sources
57 new pieces reviewed in full

↓

INDEPENDENT ARTICLE REVIEW

↓

ARTICLE REVIEW MEETINGS

↓

OUTCOME
216 sources accepted for review

Figure 1. Search strategy and results.

Though there was little dedicated focus on the patient's experience of medical tourism in the included sources, particularly that which had come from consulting directly with medical tourists themselves, when taken together they collectively contribute to answering the scoping review question. This contribution comes in the form of extracted informational points relating to the four themes identified across the sources, namely: (1) decision-making; (2) motivations; (3) risks; and (4) first-hand accounts. These four themes represent the most discussed issues about the patient's experience of medical tourism in the English-language academic, media, and grey literatures. In the remainder of this section we expand on these themes. Given the large number of sources included in the review, for manageability we cite no more than six at a time. While patients within the developing world do indeed travel abroad for health services, sometimes as medical tourists, the bulk of the sources focus on travel from developed to developing countries (referred to sometimes as north-south or west-east medical tourism) [20, 21, 22]. As such, in the remainder of the article we are referring to patients traveling from developed nations to developing ones unless otherwise stated, though certainly some of the findings are applicable to south-south patient flows.

3.1 Decision-making

Three issues were most discussed with regard to decision-making: (1) push factors (i.e., things that drove patients away from care at home); (2) pull factors (i.e., things that drew patients to other countries); and (3) information sources consulted. The push factor noted most frequently was that of cost [25, 26, 27, 28, 29, 30]. It was commonly reported that the high cost of out-of-pocket payments for procedures in patients' home countries likely pushes them to consider medical tourism [5, 17, 31,32, 33, 34]. Related to cost, a lack of insurance, or being underinsured, may also push people into medical tourism, in that if procedures cannot be covered by their insurance plans then they may be pushed abroad in search of more affordable care [5, 17, 35, 36, 37, 38]. The other most frequently noted push factor was that of wait-times, with the promise of more timely care in other countries potentially drawing them abroad [17, 39, 40, 41, 42, 43].

The pull factors identified in the reviewed sources were more numerous than the push factors. The most frequently discussed pull factor was quality. Patients can be pulled towards medical tourism hospitals that are renowned for the quality of service, care, and facilities they offer

[1, 20, 26, 32, 40,44]. Language is also factored into decision-making, wherein patients are thought to be drawn to receiving care in places where hospital employees speak their language [1, 45, 46, 47, 48]. Related to this, the religious accessibility of medical tourism facilities and destination nations is another consideration that can pull patients to one location over another. More specifically, patients may seek out facilities that observe the same religious protocols they do [48, 49]. The political climate of countries or regions may also pull people to receive care in particular locations, in that patients are unlikely to want to travel to places that are politically (or even culturally) unstable or inaccessible [20,50, 51, 52]. The vacation aspect of medical tourism serves as another pull factor, in that patients may be drawn to receive care in places they are interested in holidaying in [2, 3, 46, 53, 54, 55].

Having access to information while decision-making about medical tourism is vital given the range of factors that patients needed to consider before committing to going abroad, such as the credentials of doctors [56, 57, 58]. The international marketing of facilities and procedures, including their costs, online, via facilitators/brokers, aids some patients in decision-making [13, 27, 59, 60, 61, 62]. The presence of such marketing informs potential patients about treatment options, tourism opportunities, and other key pieces of information that assist with decision-making. The internet also offers patients other types of information about medical tourism. For example, websites created by former medical tourists to share their experiences can act as an information source for those at the decision-making stage [56]. Word-of-mouth is also an important information source, with some medical tourists having first learned about the potential for accessing procedures abroad from friends and family [6, 20, 34, 63, 64, 65].

3.2 Motivations

In addition to wanting to address a personal health need through surgery, the review captured three types of factors motivating patients' engagement in medical tourism: (1) procedure-based; (2) travel-based; and (3) cost-based. A procedure-based motivator noted in several sources is that patients may wish to pursue procedures abroad that are illegal or not available in their home countries [1,66, 67, 68, 69, 70]. For example, some patients can gain access to experimental procedures abroad that have yet to be approved for use by doctors in their local hospitals, such as stem cell therapies [71]. Related to this, medical tourism can enable access to specific expertise and specialization [44] as well as advanced technologies [22, 66, 72, 73]. These factors may also be particularly motivating for middle- and upper-class residents of developing countries who can afford to pay for more sophisticated care abroad [25]. Furthermore, hearing success stories about positive outcomes from others, whether by word-of-mouth or online, can serve as a motivator for potential medical tourists [20].

As noted in the previous section, the potential for travel and tourism is something that patients may consider when decision-making about medical tourism. It is thus not surprising that certain travel-based factors may actually serve as motivations for ultimately deciding on medical tourism, including on the destination location. Such motivators are thought to be the increasing ease and affordability of international travel, the frequency of flights to major destinations, and the streamlining of visa procedures and expediting of applications for international patients [47, 74, 75,76, 77]. The availability of facilitators/brokers to assist with making detailed arrangements, corresponding with doctors, and planning after care in certain departure countries (e.g., in Canada it is reported that there are at least 20 different facilitators/brokers) can serve as a motivator for those reluctant to have to make their own plans and bookings [6, 29, 36, 72]. Another travel-based motivator may be the presence of package deals, in that their affordability and ease of booking again may appeal to those looking for guidance in planning [45, 72, 78].

Not surprisingly, cost and affordability were often discussed as potential motivations for patients' engagement in medical tourism. In countries that have public health care coverage, such as Canada, the (often inaccurate) perception among patients that they may receive a partial or full reimbursement for the travel and procedure costs incurred abroad is thought to motivate some to engage in medical tourism [40, 73, 79, 80]. However, it was noted that restrictions were typically placed on the availability of reimbursements for planned procedures undertaken abroad via medical tourism and that out-of-pocket cost coverage is extremely rare [6, 40, 80, 81, 82]. A significant amount of the discussion coming out of the US, a country without a public health insurance plan, reveals that others' desires to keep patients' health care costs low may ultimately influence patients' choices regarding going abroad. More specifically, some employers and insurance companies are encouraging people to access surgeries abroad because costs, even when travel is factored in, are substantially lower than what would be incurred at home [1, 64, 83, 84, 85].

3.3 Risks

Given the challenges that people may face when undergoing surgery or traveling abroad, it is not surprising that the risks of medical tourism, which combines both of these aspects, were discussed throughout the reviewed literature. Three categories of risk were covered: (1) risks to patients' health; (2) risks of travel; and (3) risks pre- and post-operatively in the home country. Risks that patients may be exposed to that can have clearly negative impacts on their health include: contracting an infection post-operatively while in the hospital, travelling during the recuperative period, and an inadequate blood supply being available on-site at the hospital to meet the patients' needs [35, 86, 87]. It is also thought that undertaking procedures that are illegal in patients' home countries or experimental may expose medical tourists to unknown health risks, which may be the very reason that these same surgical procedures are not being performed in their home countries [46]. Related to cost motivations and decision-making factors, there is also a broad concern that making clinical decisions based on procedure costs is risky and may have negative outcomes for medical tourists and their health [39].

While there are always risks associated with travel, two travel-related risks that are particularly relevant to medical tourists were highlighted in the literature. The first pertains to airline travel. More specifically, flying with a serious health condition either in advance of the surgery (i.e., while getting to the hospital) or post-operatively increases patients' susceptibility to deep vein thrombosis (or sometimes referred to as 'traveler's thrombosis' when occurring as a result of a long-haul flight) [20, 36, 88, 89]. This occurs when blood clots form in deep veins and cause affected areas to swell. Patients not well enough to move around during the flight may have restricted blood flow to areas of the body, thus resulting in deep vein thrombosis [86]. In some instances the clots may dislodge and travel to the lungs, thereby causing a pulmonary embolism. It was also noted that being away from family, particularly during the recuperative period abroad, and the mental strain of travel may lead to the onset of psychological and/or emotional stress for medical tourists, thus posing as another type of travel-related health risk [5, 13, 21].

The risks of medical tourism are not restricted to occurring while patients are abroad, nor do they end upon check-out from the hospital. Patients may not seek advice from their regular doctors, or may go against their doctors' advice, regarding whether or not surgery is needed [19, 71, 90, 91]. Related to this, patients' medical records may become discontinuous, in that there are not presently adequate systems in place for transferring health information between medical tourism hospitals and patients' home doctors [39, 92, 93]. This problem may be overcome by patients carrying their records with them overseas and bringing back new files from abroad for inclusion in their permanent records [65]. Some reviewed sources further suggested that there may also be health risks upon return due to a lack of after-care planning [42, 89, 94], or that after-care may be challenging due to informational discontinuity [85]. Another risk post-operatively that may be experienced upon returning home is that there is commonly little legal recourse for procedures for which complications have arisen [86, 95, 96]. This is primarily due to weak malpractice laws that exist in many destination countries [73, 92, 97]. A related risk is that some doctors in home countries may be reluctant to treat medical tourists upon returning home for fear that they will be sued for complications arising from procedures undertaken abroad in countries with limited options for legal redress [90].

3.4 First-hand Accounts

First-hand accounts of medical tourists' experiences were found mostly in reports and media sources. These accounts typically focused on one of three things: (1) positive and negative aspects of medical tourism; (2) sensationalized issues; and (3) reports of post-recovery life. Topics often covered in these accounts ranged from reports of being satisfied with the care received [11] and the benefits and drawbacks of recovering in a relaxing tourist locale away from home [42, 98] to the experience of deciding on a particular hospital [61]. The perspectives offered on these issues were quite broad, and covered both positive and negative experiences. On the more negative side, accounts covered details such as patients having to borrow heavily from family in order to afford to access care abroad [18] and concerns about being treated by foreign doctors who may speak a different language and have different care standards [2].

One of the more sensationalized topics shared in patients' published accounts of medical tourism were stories of their own and others' expectations of cleanliness and care quality in hospitals abroad. A patient reporting on care having been received in Thailand said: "...this is not a straw-village clinic with rusty scalpels!" (p. 388) [38]; meanwhile another who accessed dental surgery in China reported: "It was dubious when you looked at it [the clinic], but when you got into the place they were competent, intelligent, and did everything they had to do" (p. 68) [99]. It is not surprising that patients emphasized such issues in their accounts given that there were reports of having to counter others' perceptions of care abroad. As a woman from the US who had received surgery in Thailand explained: "They [friends, family, and others] roll their eyes up in their heads and say, 'I can imagine' and I say, 'no, you can't...I went down and had lunch at the Starbucks in the lobby of the hospital, came back up and the doctor had on his desk the most beautiful file, all bound with tabs and everything, with all the results of the tests that they had done'... Something like that, as you know, is impossible in America...I mean, it's inconceivable" (p. A6) [100]. Others shared this experience, including another US patient who said that: "when I told people I was having surgery in Southeast Asia, some looked at me like I was crazy. They were clearly imagining me in a straw hut with someone holding fishing line and tweezers" (p. P01) [101]. Others' accounts focused on how care at hospitals abroad was not as sterile as they had expected [102].

Retrospective, post-recovery reports were abundant in the media sources reviewed. In these accounts former medical tourists reflect, overwhelmingly positively, on their experience overall and the positive impact that receiving a procedure abroad has had for their health. A former Canadian medical tourist had this to say: "Life is too precious. I'm in my early 50s and I have lots of things to do in my life and one of them isn't lying at home in pain... I'm a Victoria firefighter. I have been for 29 years and I don't want to retire that way, you know, with a disability" (n.p.) [103]. Accessing surgery abroad enabled him to maintain his employment. Another Canadian retrospectively proclaimed "I think it's the best money I've ever spent" (n.p.) [104], with regard to having paid out-of-pocket for surgery in India.

4. DISCUSSION

Table 3 summarizes the themes and issues generated from the 216 sources included in this scoping review. The informational points extracted in the review, generated primarily from sources reliant on speculation, reveal how complex the patient experience is, in that it involves many components ranging from early decision-making about multiple factors to ensuring appropriate arrangements for post-operative care in one's home system. The informational points also suggest that multiple individuals are involved in shaping the patient experience, including: family and friends who may have shared their own successful experiences of receiving care abroad with potential medical tourists; surgeons, nurses and patient coordinators who have direct responsibility for how on-site care is delivered in destination hospitals; and medical tourism facilitators/brokers who work to assist patients at multiple points. In the remainder of this section we explore the implications of the key issues identified about the patient's experience of medical tourism, knowledge gaps that exist on this topic, and also the overall limitations of the scoping review process employed.

4.1 Implications for Patients

From the sources reviewed for this scoping review it can be understood that patients are likely to hold significant responsibilities in the practice of medical tourism. For example, informational continuity of care is a quality indicator and is established through patient information being available over time and to multiple practitioners in different locations [105, 106]. While patients often have roles to play in its establishment [107], medical tourists may hold particular responsibilities in this regard as they may literally be expected to transport hard copy records vast distances and ensure that they arrive safely to the correct people. Certainly, concerns are that these records may become damaged during transit or that the patient may choose not to share details of their procedure abroad with their regular doctor, thereby threatening informational continuity of care and the benefits it bestows.

Another responsibility is that it may be advisable for patients to take active measures to avoid encountering risks when traveling and also while abroad. While all patients run the risk of being exposed to any number of health threats when receiving surgery, such as surgical site infections and clotting complications [108, 109], there can be additional risks that pertain to the 'travel' and 'receiving care abroad' dimensions of medical tourism that patients must take responsibility for minimizing or eliminating. Related to this, patients interested in medical tourism are also likely to hold responsibility for evaluating the trustworthiness and reliability of information sources (e.g., promotional materials, facilitators/brokers, friends and family). It has been said that international regulation of the medical tourism industry is lacking [110], and so patients are left to their own - possibly with the assistance of others - to rate and rank things such as the quality of facilities and procedure outcomes using available information. Clearly, it can be understood that how patients address these potential responsibilities is likely to directly shape their experiences of intentionally accessing medical care abroad via medical tourism.

Table 3. Summary of extracted informational points

Themes Identified	Main Issues Covered in Reviewed Sources	Example
Decision-making	Push factors	High out-of-pocket costs for procedures in patients' home countries
	Pull Factors	Hospitals known to deliver high quality care
	Information Sources Consulted	Word-of-mouth
Motivations	Procedure-based motivations	Wanting access to procedures that are illegal or unavailable in home country
	Travel-based motivations	Ease of booking
	Cost-based motivations	Recommendation by employer or insurance company as a cost savings measure
Risks	Risks to health	Contracting infection while abroad
	Risks of travel	Flying post-operatively
	Pre- and post-operative risks at home	Little legal recourse in certain jurisdictions
First-hand accounts	Positive and negative aspects of experiences	Reports of being satisfied with care received
	Sensationalized issues	Reports of needing to counter others' negative perceptions about destination countries
	Post-recovery life	Reports of improved health status

This scoping review has revealed that the patient's experience of medical tourism might not end upon returning home, nor begin at the point of departure. In fact, it can extend far in advance of and beyond when care is received. The law reviews examined have focused extensively on the fact that there is typically little legal recourse for patients accessing surgeries and other procedures abroad in countries with weak malpractice laws. For patients who experience negative outcomes, their pursuit of compensation, whether financial or otherwise, may draw on over an extended period of time, quite possibly with little result, extending well beyond the post-operative recovery period.

It was noted in the findings section that word-of-mouth is likely to be important within the medical tourism industry: satisfied patients can spread information about facilities and destination countries to interested others. In this way, these medical tourists may become 'ambassadors' for destination countries and hospitals over time after returning home. In doing so they may effectively challenge perceptions of the health care delivered in some countries, thus working to overcome the kinds of 'straw hut' perceptions of care shared in the first-hand accounts subsection. At the same time, negative experiences of care abroad may serve to further entrench these perceptions.

4.2 Knowledge Gaps

Although a significant number of sources were accepted into this scoping review, the vast majority did not present empirical research or other tested facts. Instead, this review has revealed that most of what is known about the patient's experience of medical tourism is, in fact, speculative, idea-based, or anecdotal in nature. The frequency with which some things were reported in the sources reviewed, such as cost savings as being a significant motivator for patient engagement and waiting lists pushing patients abroad, suggests that there is some consensus about specific aspects of the patient experience. However, often authors looked to media sources to verify such aspects given that empirical evidence is lacking. This is not surprising as newspaper and magazine articles were found to be some of the only sources that shared first-hand accounts from medical tourists, thus allowing for limited confirmation about ideas regarding motivations and other factors central to the patient's experience. Put another way, the area of medical tourism is ripe for research to be conducted by social and health scientists alike from a range of disciplinary and methodological perspectives. Research is needed not only to confirm or refute long-standing speculation regarding the patient's experience, but also to 'shed light' on areas that have been given almost no consideration to-date. For example, it seems that little attention has been paid to establishing the size and directions of the international flow of patients or the success rates of procedures performed at medical tourism hospitals. No doubt data access is a challenge, in that private hospitals treating international patients are likely reluctant to share such information.

A number of specific knowledge gaps are evident within the four thematic areas regarding the patient's experience of medical tourism identified in the scoping review. Research attention needs to be given to understanding how information sources consulted and evaluated by patients prior to departure. It would also be useful to better understand how patients understand the risks of accessing care abroad at this point in time. Retrospective accounts upon returning home could then shed light onto whether or not important factors were adequately considered in advance of ultimately deciding on accessing care abroad. Such research could offer useful insights into what patients need to know before going abroad and could ultimately assist with developing patient-focused decision-making aids. Significant knowledge gaps also exist regarding the push factors, pull factors, and motivations identified in the scoping review. Research needs to be conducted to confirm that those factors cited by the sources reviewed are indeed accurate, and also to determine whether or not others exist and also how they might differ between individuals and also patient groups (e.g., by procedure type, by home country, by destination country). In fact, any research that contributes to enhancing our understanding of how experiences of medical tourism differ between patients and the roles of factors such as socio-economic status, diagnosis, and overall health status in such differentiation is highly needed.

It was noted at the outset of this section that multiple individuals and groups are involved in shaping the patient's experience of medical tourism. As such, it is essential to turn research attention to these important stakeholders so as to better contextualize how patients experience this practice. This could include investigating doctors' and facilitators/brokers' roles in assisting patients during the decision-making process and also in sharing information on the risks. The practice of medical tourism holds significant implications for the airlines that transport patients and also travel agents who do bookings for patients not using the services of facilitators/brokers. There was almost no discussion of these important stakeholders in the sources reviewed in this scoping review. An understanding of the perspectives of these businesses and others who are linked to the industry (e.g., travel insurance, hotels) on the soundness of the practice of medical tourism and the ways in which it strengthens and/or threatens their operations is needed. This research would also be valuable because so very little consideration has been given to the tourism aspect of medical tourism. Because of this oversight, little is known about the importance of the tourism aspects of travel and decision-making, among other factors, for medical tourists. Tourism scholars, thus, have the ability to make important contributions to this area of health services knowledge.

5. LIMITATIONS

A comprehensive approach was taken to scoping a variety of sources to synthesize what is known about the patient's experience of medical tourism. The inclusion of steps such as having two reviewers for every full source, developing a search strategy in consultation with a librarian, and searching for sources of all types from a comprehensive grouping of databases have added rigour to the scoping process and thus serve as strengths. However, two main limitations exist. The first is that only English-language sources were retrieved and reviewed. No doubt there is literature on medical tourism that has been produced in other languages. At the same time, there were very little non-English-language sources cited in the pieces reviewed, and so this suggests that the most important sources may be available in English. For example, industry and government reports produced in non-English-speaking countries were commonly available in English and so were included in the review process (e.g., [25]). The second main limitation is that the media sources reviewed were limited to Canada and a few major North American magazines and newspapers known to cover Canadian health and health care issues. Placing this restriction on the inclusion of media sources was necessary in order to keep the review manageable. As such, the media sources included are presented as a representation of the types of local, regional, and national coverage that exists of medical tourism within a particular country known as a departure point for medical tourists.

6. CONCLUSIONS

Given the burgeoning, complex, and frequently controversial phenomena associated with medical tourism, this article has presented a much-needed scoping review to illustrate what is known (and not known) about this practice, especially in terms of patients' experiences. Addressing multiple types of literature, we hope this comprehensive knowledge synthesis will usefully guide research, government, and industry agendas alike. Seeking a focused understanding of medical tourism, full review of the 216 included sources (the overwhelming majority of which did not present empirical findings) identified four main themes that summarize what is known about the patient's experience. These themes are characterized by a focus on: (1) patients' *decision-making* in terms of push and pull factors, as well as the consultation of information; (2) *motivations* related to procedure, travel, and cost; (3) *risks* associated with patients' health, travel, and pre- and post-operative conditions in the home country; and (4) *first-hand accounts* of the positive and negative components of medical tourism, sensationalized issues, and post-recovery life. Using insights gleaned from the scoping review, we suggest that medical tourism is likely to have significant implications for patients, including that patients may very well hold a number of responsibilities should they choose to engage in intentionally going abroad for non-emergency care that is paid for out-of-pocket. Among these is potentially playing a role in ensuring that continuity of care is maintained as best as possible despite the disjuncture in their care trajectory across countries and providers.

Despite engagement with the issue of medical tourism in the published literature, it is clear that there is profound paucity of theoretical and empirical understandings of this practice that can ultimately help us to understand the patient's experience. With empirical evidence lacking, authors have frequently drawn on media sources to substantiate their claims. We believe that the time is ripe for social and health scientists from various disciplinary, theoretical, and methodological perspectives to go beyond enduring speculations about patients' experiences of medical tourism. As a starting point, we have identified a number of pressing research direction above. Finally, we hope the identified knowledge gaps and research challenges, along with the scoping review findings, illustrate the exciting possibilities for how scholars can make significant contributions to what is known about the patient's experience of medical tourism.

ACKNOWLEDGEMENTS

Funding was provided by a Catalyst Grant from the Canadian Institutes of Health Research. Thanks go to librarian Heather DeForest for consulting with the team on the search strategy.

AUTHORS' CONTRIBUTIONS

VAC led the identification of themes and writing of this manuscript. All authors contributed to the design of the scoping review, reviewing sources, and assisting with data management and interpretation. All authors provided feedback throughout the drafting of this article and have read and approved the final manuscript.

REFERENCES

1. Burkett L: Medical tourism: concerns, benefits and the American legal perspective. J L Med. 2007, 28: 223-245.
2. Mudur G: Hospitals in India try and woo patients. BMJ. 2004, 328: 1338-1338.
3. Ramirez de Arellano A: Patients without borders: the emergence of medical tourism. Int J Health Serv. 2007, 37: 193-198.
4. Unti JA: Medical and surgical tourism: the new world of health care globalization and what it means for the practicing surgeon. Bull Am Coll Surg. 2009, 94: 18-25.
5. Howze KS: Medical tourism: symptom or cure?. Georgia L Rev. 2007, 41: 1013-1052.
6. Klaus M: Outsourcing vital operations: what if US health care costs drive patients overseas for surgery?. Quinnipiac Health L J. 2005, 9: 219-247.
7. Turner L: Canadian Medicare and the Global Health Care Bazaar. Policy Options. 2007, C: 73-77.

8. Chanda R: Trade in Health Services. Trade in Health Services: Global, Regional and Country Perspective. Edited by: Drager N. 2002, Washington DC: Pan-American Health Organization, 35-44.

9. Pennings G: Reproductive tourism as moral pluralism in motion. J Med Ethics. 2002, 28: 338-341.

10. Bookman MZ, Bookman KR: Medical Tourism in Developing Countries. 2007, New York, NY: Palgrave Macmillan

11. Ehrbeck T, Guevara C, Mango PD, Cordina R, Singhal S: Health care and the consumer. McKinsey Quarterly. 2008, 4: 80-91.

12. Keckley PH, Underwood HR: Medical Tourism: Update and Implications. 2008, Washington, D.C.: Deloitte Centre for Health Solutions

13. Law J: Sun, sand and stitches. Profit. 2008, 27: 69-70.

14. Arksey H, O'Malley L: Scoping studies: Towards a methodological framework. International Journal of Social Research Methodology. 2005, 8: 19-32.

15. Pachanee C, Wibulpolprasert S: Incoherent policies on universal coverage of health insurance and promotion of international trade in health services in Thailand. Health Policy Plan. 2006, 21: 310-318.

16. Bertinato L, Busse R, Fahy N, Legido-Quigley H, McKee M, Palm W, Passarani I, Ronfini F: Policy Brief Cross-Border Health Care in Europe. 2005, European Observatory on Health Systems and Policies: World Health Organization

17. Garcia-Altes A: The development of health tourism services. Ann Tourism Res. 2005, 32: 262-266.

18. Marlowe J, Sullivan P: Medical Tourism: The Ultimate Outsourcing. Human Resources Planning. 2007, 30 (2): 8-10.

19. Whittaker A: Pleasure and pain: medical travel in Asia. Global Pub Health. 2008, 3: 271-290.

20. Kangas B: Hope from abroad in the international medical travel of Yemeni patients. Anthro Med. 2007, 14: 293-305.

21. Lautier M: Export of health services from developing countries: The case of Tunisia. Soc Sci Med. 2008, 67: 101-110.

22. Kangas B: Therapeutic itineraries in a global world: Yemenis and their search for biomedical treatment abroad. Med Anthro. 2002, 21: 35-78.

23. Hanna SA, Saksena J, Legge S, Ware HE: Sending NHS patients for operations abroad: is the holiday over?. Ann R Coll Surg Engl. 2009, 91: 128-130.

24. Wibulpolprasert S, Pachanee C, Pitayarangsarit S, Hempisut P: International service trade and its implications for human resources for health: a case study of Thailand. Human Res Health. 2004

25. Leng CH: Medical tourism in Malaysia: international movement of healthcare consumers and the commodification of healthcare. Asia Res Inst. 2007, 1-32. ART WP No. 83

26. Horowitz MD, Rosensweig JA: Medical tourism: health care in the global economy. Phys Exec. 2007, 33: 24-30.

27. Moore A: Health tourism: don't forget your toothbrush... Health Serv J. 2009, 119: 18-20.

28. Garud A: Medical tourism and its impact on our healthcare. Natl Med J India. 2005, 18: 318-319.

29. Lindsay G: Medical leave. Fast Comp. 2008, 125: 108-119.

30. Medical tourism for insured patients at a glance. 2009, Assoc Press, 312.

31. Crone RK: Flat medicine? exploring trends in the globalization of health care. Acad Med. 2008, 83: 117-117.

32. Gray HH, Poland SC: Medical tourism: crossing borders to access health care. Kennedy Inst Ethics J. 2008, 18: 193-201.

33. Johnson LA: Americans looking near, far to save on health care. 2008, Assoc Press

34. Turner L: Cross border dental care: 'dental tourism' and patient mobility. Br Dent J. 2008, 204: 553-554.

35. Forgione DA, Smith PC: Medical tourism and its impact on the US health care system. J Health Care Fin. 2007, 34: 27-35.

36. Herrick D: Medical tourism: global competition in health care. Natl Cnt Pol Analysis. 2007, 1-40. NCPA No. 304.

37. Richards T: Border crossing. The medical travellers' tale: so far health tourism is a small market, but it is set to stimulate welcome competition. BMJ. 2008, 337: 1324-1324.

38. Operating profit; Globalisation and health care. Economist. 2008, 388.

39. Cheung IA, Wilson A: Arthroplasty tourism. Med J Aust. 2007, 187: 666-667.

40. Eggerston L: Wait-list weary Canadians seek treatment abroad. CMAJ. 2006, 174: 1247-1247.

41. Leahy AL: Medical Tourism: the impact of travel to foreign countries for healthcare. Surgeon. 2008, 6 (5): 260-261.

42. Choat I: Surgically enhance your wage. Travel Weekly. 2003, 1688: 38-39.

43. Neelakantan S: India's global ambitions. Far East Eco Rev. 2003

44. Hume L, DiMicco FJ: Bringing hotels to healthcare: a RX for success. J Quality Assurance Hospitality Tourism. 2007, 8: 75-84.

45. Saniotis A: Changing ethics in medical practice: a Thai perspective. Ind J Med Ethics. 2007, 4: 24-25.

46. Sen Gupta A: Medical tourism in India: winners and losers. Ind J Med Ethics. 2008, 5: 4-5.

47. Lee C: Just what the doctor ordered: medical tourism. Monash Bus Rev. 2007, 3: 10-12.

48. Connell J: Medical tourism: sea, sun, sand and... surgery. Tourism Mgmt. 2005, 27: 1093-1100.

49. After 9/11, Arabs turning to Asia and away from US for medical care. 2006, Assoc Press

50. Smith PC, Forgione DA: Global outsourcing of health care: a medical tourism decision model. JITCAR. 2008, 9: 19-30.

51. Colias M: Circling the globe. Hosp Health Networks. 2004, 78: 14-14.

52. Rhea S: They're coming to America ... Mod Healthc. 2007, 37: 20-20.

53. Goodrich J: Socialist Cuba: a study of health tourism. J Travel Res. 1993, 32: 36-41.

54. 'Medical tourism' on the rise. Advert Age. 2005, 76: 14-14.

55. Sen Gupta A: Medical tourism and public health. People's Dem. 2004, 28:

56. Badam RT: Americans, Europeans head to India for cheap, high-quality medical care. Guardian. 2005, B7.

57. Perotin MM: Sun, sea and surgery. Natl Post. 2004, PT3.

58. Olian C: 60 minutes: Medical Tourism. CBS Television. 2006,

59. Reed CM: Medical tourism. Med Clin N Am. 2008, 92: 1433-1446.

60. Blesch G: Rising in the Persian Gulf. Dubai and other United Arab Emirates states are using U.S. know-how to build a cutting-edge healthcare infrastructure. Mod Healthc. 2007, 37: 26-28.

61. Cortez NG: Patients without borders: the emerging global market for patients and the evolution of modern health care. Indiana LJ. 2008, 83: 71-132.

62. Horowitz MD, Rosensweig SA, Jones CA: Medical tourism: globalization of the healthcare market. Med Gen Med. 2007, 9: 33-33.

63. Davolt S: More U.S. patients boldly go where medical tourists have gone before. Emp Benefit News. 2006, 20: 1-82.

64. Butler S: Holidays for health; people are traveling overseas for cheaper medical procedures like colonoscopies and hip replacements. Newsweek Intl Edition Spec Rep Travel. 2009

65. Wolff J: Passport to CHEAPER HEALTH CARE?. Good Housekeeping 1. 2007, 245: 190-190.

66. DiMicco F, Centron M: Club Medic. Asian Pac Biotech News. 2006, 10: 527-531.

67. Dunn P: Medical tourism takes flight. Hosp Health Networks. 2007, 81: 40-44.

68. Jesitus J: Safari surgery. Cosmetic Surg Times. 2006, 9: 1-14.

69. Medical tourism soars in popularity. J Hosp Palliat Nurs. 2007, 9: 234-234.

70. Werb J: Sun, sea and surgery. BC Bus. 2007, 35: 17-17.

71. Vastig B: Unproven treatments. US News World Reports. 2008, 144: 50-50.

72. Oberholzer-Gee F, Khanna T, Knoop CI: Apollo-hospitals--first-world health care at emerging-market prices. Harv Bus Sch. 2005, HBS No. N9-705-442.

73. Medical tourism: need surgery, will travel. CBC News Online. 2004, [http://www.cbc.ca/news/background/healthcare/medicaltourism.html]

74. Jenner EA: Unsettled borders of care: medical tourism as a new dimension in America's health care crisis. Res Soc Health Care. 2008, 26: 235-249.

75. Bajaj K: Come, Heal Thyself. Business Today (India). 2007

76. Chinai R, Goswami R: Medical visas mark growth of Indian medical tourism. Bull World Health Org. 2007, 85: 164-165.

77. Arunandondchai J, Fink C: Globalization for health: trade in health services in the ASEAN region. Health Promo Int. 2007, 21: 59-66. 10.1093/heapro/dal052.

78. Turner L: Commercial Organ Transplantation in the Philippines. Cambridge Quarterly of Healthcare Ethics. 2009, 18: 192-196.

79. Prashad S: The world is your hospital. Can Bus. 2008, 81: 62-66.

80. Starnes R: Frustrated by waiting lists, Canadians help fuel medical tourism boom abroad. CanWest News. 2004, 1-1.

81. City resident opts for hip surgery overseas. Burnaby Now. 2008, 5.

82. CMA calls for national solution to medical tourism. CBC News. 2007, 11:39 AM GMT

83. Marlow J, Sullivan P: Medical tourism: the ultimate outsourcing. Hum Resour Plan. 2007, 30: 8-10.

84. U.S. options available for medical tourists. Bus Insur. 2008, 42: 8-8.

85. Importing Competition: globalisation and health. Economist. 2008, 388: 12-12.

86. MT: Traveller's dream or healthcare nightmare?. Nurs Pract (Lond). 2007, 38: 4-4.

87. McQueen MP: Paying workers to go abroad for health care. WSJ. 2008, 252: B9.

88. Reese S: CARE beyond borders. Managed Healthc Ex. 2007, 17: 33-36.

89. Carabello J: A medical tourism primer for U.S. physicians. J Med Pract Mgmt. 2008, 23: 291-291.

90. Brouwer W, van Excel J, Hermans J, Stoop A: Should I stay or should I go? waiting lists and cross-border care in the Netherlands. Health Pol. 2003, 63: 289-298.

91. Balfour F, Kripalani M, Capell K, Cohn L: Sand, sun and surgery. Bus Week. 2004, 3870: 48-49.

92. Canada AM: More Canadians travelling abroad for medical care. 2007, CTV Television Inc

93. Canada AM: Medical tourism: Reaction from health-care establishment. 2007, CTV Television Inc

94. Turner L: Medical tourism: family medicine and international health-related travel. Can Fam Phys. 2007, 53: 1639-1641.

95. Shimo A: New hot spots for surgery. Maclean's. 2000, 120: 44-44.

96. Smerd J: A ticket to lower care costs. Workforce Mgmt. 2006, 85: 1-33.

97. Milstein A, Smith M: America's new refugees - seeking affordable surgery offshore. N Engl J Med. 2006, 355: 1637-1640.

98. Nip, tuck vacations risky, doctors warn. 2007, CBC News, [http://www.cbc.ca/consumer/story/2007/01/10/cosmetic-surgery.html]

99. Patriquin M: A vacation that'll have you drooling. Maclean's. 2007, 120: 68.

100. Mydans S: The Perfect Thai Vacation: Sun, Sea and Surgery. NYT. 2002, A6-[http://www.nytimes.com/2002/09/09/world/the-perfect-thai-vacation-sun-sea-and-surgery.html?scp=1&sq=The%20Perfect%20Thai%20Vacation:%20Sun,%20Sea%20and%20Surgery.%20%20&st=cse]

101. Loose C: Operation vacation; big savings have more overseas travelers mixing surgery with sightseeing. Wash Post. 2007, P01-Travel:.

102. Hewitt B, Huston A, Dhillon A, Haederle M: The Doctor is in... INDIA. People. 2006, 65: 131-135.

103. Copeland J: Canadian medical tourist in India. 2004, CBC News, [http://www.cbc.ca/news/viewpoint/vp_copeland/20040920.html]

104. Termonti AM: The Current. CBC Television. 2005, [http://archives.cbc.ca/programs/719/]

105. Donaldson MS: Continuity of Care: A Reconceptualization. Medical Care Research and Review. 2001, 58 (3): 255-290.

106. Saultz JW, Albedaiwi W: Interpersonal continuity of care and patient satisfaction: a critical review. Annals of Family Medicine. 2004, 2 (5): 445-451.

107. Crooks VA, Agarwal G: What are the roles involved in establishing and maintaining informational continuity of care within family practice? A systematic review. BMC Family Practice. 2008, 9: 65-10

108. Graves N, Halton K, Doidge S, Clements A, Lairson D, Whitby M: Who bears the cost of healthcare-acquired surgical site infection?. Journal of Hospital Infection. 2008, 69: 274-282.

109. Ferraro G, Grella R, D'Andrea F: Abdominoplasty: Thromboembolic risks for both sexes. Aesthetic Plastic Surgery. 2004, 28 (6): 412-416.

110. Terry NP: Under-regulated health care phenomena in a flat world: Medical tourism and outsourcing. Western New England Law. 26: 421-426.

Chapter 2

Promoting social responsibility amongst health care users: medical tourists' perspectives on an information sheet regarding ethical concerns in medical tourism

Krystyna Adams[1*], Jeremy Snyder[1], Valorie A Crooks[2] and Rory Johnston[2]

[1]Faculty of Health Sciences, Simon Fraser University, Blusson Hall, 8888 University Drive, Burnaby, British Columbia, Canada.

[2]Department of Geography, Simon Fraser University, Robert C. Brown Building, 8888 University Drive, Burnaby, British Columbia, Canada.

ABSTRACT

Background

Medical tourists, persons that travel across international borders with the intention to access non-emergency medical care, may not be adequately informed of safety and ethical concerns related to the practice of medical tourism. Researchers indicate that the sources of information frequently used by medical tourists during their decision-making process may be biased and/or lack comprehensive information regarding individual safety and treatment outcomes, as well as potential impacts of the medical tourism industry on third parties. This paper explores the feedback from former Canadian medical tourists regarding the use of an information sheet to address this knowledge gap and raise awareness of the safety and ethical concerns related to medical tourism.

Results

According to feedback provided in interviews with former Canadian medical tourists, the majority of participants responded positively to the information sheet and indicated that this document prompted them to engage in further consideration of these issues. Participants indicated some frustration after reading the information sheet regarding a lack of know-how in terms of learning more about the concerns discussed in the document and changing their decision-making. This frustration was due to participants' desperation for medical care, a topic which participants frequently discussed regarding ethical concerns related to health care provision.

Conclusions

The overall perceptions of former medical tourists indicate that an information sheet may promote further consideration of ethical concerns of medical tourism. However, given that these interviews were performed with former medical tourists, it remains unknown whether such a document might impact upon the decision-making of prospective medical tourists. Furthermore, participants indicated a need for an additional tool such as a website for continued discussion about these concerns. As such, along with dissemination of the information sheet, future research implications should include the development of a website for ongoing discussion that could contribute to a raised awareness of these concerns and potentially increase social responsibility in the medical tourism industry.

1. BACKGROUND

Medical tourism, the practice of patients traveling out of country with the intention to receive medical care paid for out of pocket, is considered an expanding industry globally [1, 2]. While there is a dearth of empirical evidence on patient flows and the impacts of medical tourism, media coverage often provides anecdotal stories that indicate the potential for surgical complications, and poor quality or unnecessary care due to a lack of regulation in the industry [3]. Partially due to the lack of empirical evidence available about medical tourism, researchers are concerned about the gaps in the information medical tourists may be accessing regarding potential impacts of medical tourism on both individual safety and health equity [4, 5].

Medical tourism is receiving increasing attention, both as a developing industry and a topic of research interest [6]. While there appears to be a wide variety of medical tourism flows, including travel between both countries of similar and different levels of economic development, information available to medical tourists demonstrates a concerted effort on behalf of low and middle income countries to entice persons to travel from developed to developing countries for medical care [7, 8]. This recent trend of travel from more developed to developing countries has become a focus for research on the impacts of medical tourism on global health equity [9]. Discussions by researchers about growing health inequities globally have identified the growth of the medical tourism industry as potentially one cause of these inequities [10]. The rise of neoliberalism and economic globalization in the past few decades has resulted in health care being increasingly conceptualized as a tradable commodity across borders. The medical tourism industry markets to patients as consumers of health care, encouraging individuals to take responsibility for their health by taking advantage of potential cost savings of medical treatment in developing countries and high quality care available in private facilities catering to medical tourists [11]. Patients may be motivated to travel to access procedures which are domestically unavailable or require undesired waiting times, receive care and recover in a relaxing or exotic landscape, and/or reduce costs for procedures domestically paid for out-of-pocket [7].

Research indicates that many Canadian medical tourists use word-of-mouth or the internet to inform their decisions regarding this practice, which may result in decision-making that is uninformed or biased around individual safety and treatment outcomes [12]. Canadian medical tourists may also be unaware of ethical concerns regarding potential impacts of medical tourism on global health equity, particularly potential impacts to destination and departure country health resources and health care systems [13, 14]. While researchers are increasingly discussing these potential ethical concerns related to medical tourism, medical tourists may be unable to act in a socially responsible manner if uninformed of these impacts [15].

Here we explore the responses of Canadian medical tourists to an information tool intended to encourage more informed and ethical decision-making around medical tourism. This tool was developed through an iterative process that drew on values from guidelines in related domains as well as public health communication research to effectively inform Canadians about ethical concerns related to medical tourism. Through analyzing feedback received in one-on-one interviews with former Canadian medical tourists, in this paper we examine the potential usefulness and impacts of such a tool on patients' decision-making, as well as the contexts in which such a tool might better inform and promote social responsibility amongst Canadian medical tourists and ultimately a more ethical medical tourism. Furthermore, this research provides insight into considerations for the field of public health in promoting social responsibility in the provision and utilization of health care.

2. ETHICAL ISSUES IN MEDICAL TOURISM AND THE NEED FOR ENHANCED AWARENESS

Despite a lack of empirical evidence on the impacts of medical tourism, researchers have indicated several ethical concerns relating to medical tourism's potential impact on both an individual and societal level within destination and departure countries, as well as on a global scale [13, 14]. Medical tourism may negatively impact upon individual health due to a lack of risk communication to patients and/or inadequate informed consent [2, 14]. Furthermore, the medical tourism industry may divert resources from the public to private sector. Increasing privatization of health care may reorient priorities in the provision of health care to provide more profitable specialized care, including care which meets the needs of foreign patients travelling for some form of specialized care [14, 16]. This reorientation of health care may reduce access to desired health care for local patients in medical tourism destination communities [1]. In countries of origin for medical tourists, potential complications affiliated with procedures obtained abroad, particularly due to a lack of regulation in the industry, may divert resources to treat these complications. Furthermore, reduced pressure for system reform if patients leave the health care system to seek out care abroad may result in only those persons able to travel as a medical tourist accessing health care [1, 13]. Finally, increased movement of patients across borders to access medical care may increase the transmission of infectious diseases [17]. Concerns for medical tourism's negative impacts on health outcomes and health equity indicate a burden of responsibility for those engaging in medical tourism to consider these impacts and act in a socially responsible manner when providing and utilizing health care [14].

The primary ethical concerns surrounding the development of medical tourism and its impacts are also related to larger impacts on global health equity. While countries that are promoting themselves as medical tourism destinations cite economic motivations for industry growth [18], these economic advantages may come at the cost of undermining initiatives that intend to improve health equity both within and between nations. Increases in privatized health care globally have detracted from health care system strengthening required to meet the goals of "Health for All" outlined in the World Health Organization's Declaration of Alma Ata [19] and protect the human right to health [20]. While some medical tourism stakeholders have mentioned potential benefits of the practice to health outcomes and health equity, including improved quality control due to international accreditation of medical tourism facilities, increased training opportunities for health professionals treating foreign patients, and health worker retention in places experiencing high levels of brain drain to other countries, lack of regulation by some destination governments and the industry itself may neglect considerations for health outcomes and health equity in stakeholder decision-making [9].

Ethical concerns for medical tourists' destination and departure countries are related to the provision of health care within national borders. For destination countries, concerns about the growth of the medical tourism sector are related to the impacts of medical tourism on public health care due to the potential growth of the private sector [8]. The promotion of medical care to foreign patients may encourage a shift in health resource allocation resulting in a potential brain drain of health human resources from the public to the private sector and the diversion of public resources such as land or public finances to medical tourism businesses [21]. Medical tourism may incentivize training of health workers for more curative, costly and complicated procedures that are appealing to medical tourists [7]. This emphasis on more curative care may result in neglect for the provision of appropriate primary health care to local populations and an increasing normalization of privatized health care [1]. For departure countries, medical tourism raises concerns for these countries' health systems due to the potential diversion of resources to treat medical complications resulting from patients engaging in medical tourism [14]. Furthermore, medical tourism raises concerns for health equity within nations if only patients willing and able to travel for medical care are able to access certain medical procedures [1].

Ethical concerns regarding patient health in medical tourism are particularly focused on the lack of neutral information provided to medical tourists regarding risks and safety concerns and the implications of this for patients' abilities to achieve informed consent. These safety concerns include potential medical complications related to the procedure itself, potential medical complications from travelling following a medical procedure, and/or inadequate continuity of care after returning home [17]. Patients may be unaware of these safety concerns due to lacking familiarity with the destination health care system, language or cultural barriers to adequate communication between medical tourists and health care workers, and/or lacking transparency regarding quality of medical facilities or health care workers [2].

Medical tourists and stakeholders in the medical tourism industry may be unaware of the ethical concerns outlined above, and as such, are unable to act in a socially responsible manner [14]. This lack of awareness may be due to biased sources of information, including sources with a vested interest in the profitability of the medical tourism industry [22]. Socially responsible engagement in the medical tourism industry requires individuals to act in a manner that respects the inherent dignity of all other humans [23]. Individuals are thus responsible for encouraging the development of social sustainability, defined as the creation of conditions that enable people to lead lives of personal value [24]. Improved health outcomes and health equity are increasingly recognized as cornerstones of social sustainability [25]. As such, the provision and utilization of health care contributes to social sustainability and necessitates responsible allocation and use of health resources in such a manner that optimizes health outcomes and health equity [26], even in relation to the private medical tourism industry. We contend that creating awareness of the ethical concerns of medical tourism will encourage socially responsible actions by individuals engaging in medical tourism, stakeholders in the medical tourism industry, and policy makers developing regulations for the industry [3]. It is for this reason that we have made an informational tool for Canadians considering medical tourism that prompts consideration of ethical, equity, and safety issues alike.

Informing individuals such as medical tourists about health issues can be done most effectively following extensive formative research to better understand the target audience and determine the ideal means of providing the information [27]. Thus, in the remainder of this article we present the findings of the formative, qualitative research we conducted in order to inform the development of an information tool for Canadians considering medical tourism that prompts consideration of ethical, equity, and safety issues. The process of formative research is useful for tailoring message points and determining effective dissemination of the information. Furthermore, formative research helps identify gatekeepers to the primary audience who either encourage or discourage information being provided to the primary audience [27]. Formative research activities such as focus groups and pre-testing of informational tools with primary target audiences and gatekeepers provide valuable insights that can be used for tool refinement and implementation [27, 28]. With this in mind, in the sections that follow we examine feedback provided to us by intended users of the informational tool so as to assess the potential impact of such an information tool, inform revision and improvement upon the tool, and determine future steps for effective dissemination of the tool leading to more informed decision-making by Canadians considering medical tourism.

3. METHODS

To develop the information tool on which the current formative study is based, an iterative process was utilized. This multi-step process is summarized in Figure 1 below. Here we report exclusively on the findings from the interviews with former Canadian medical tourists in Step 3 of this process. Although this tool was developed using ethical values and principles as a way to enhance consideration of ethical issues during the decision-making process, words such as 'ethics' and 'values' were deliberately excluded from the text as our previous research has shown that medical tourists can respond negatively to such terms, interpreting them to be explicitly judgmental. Instead, ethical concerns, along with equity and safety concerns were shared throughout the tool without being framed as such. As a health communication tool with the objective of informing and not guiding individual action, the target audience includes all Canadians that may be in the pre-contemplation or contemplation stage of decision-making regarding accessing care out of Canada as a medical tourist [29]. While this tool may provide useful information to all Canadians regarding this growing industry, it is intended to encourage more informed decision-making when considering or planning to travel out of country for private medical care, and as such, this stage of formative research focused on gaining feedback through semi-structured interviews with Canadians that have already travelled as medical tourists.

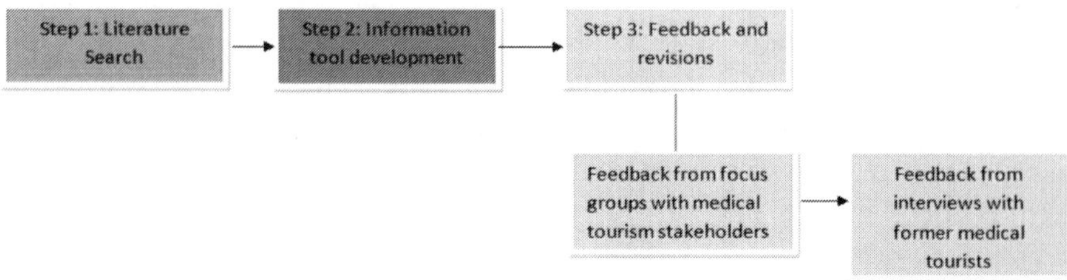

Figure 1. Multi-step process for information tool development.

3.1 Recruitment

Former medical tourists were selected to participate in this formative research as both a convenient sample to access in comparison to Canadians that may be considering participating in medical tourism, as well as a rich source of information regarding the impact of such a tool during the decision-making process given their experience. Participants were recruited using the following channels simultaneously: 1) Craigslist advertisements posted on sites for all major Canadian cities; 2) advertisement in a Vancouver newspaper; 3) posting invitations to participate in online medical tourism forums; and 4) snowball sampling through interview and focus group participants' networks. Information was provided for interested persons to contact either a toll-free phone number or email for further information and assessment of eligibility. Eligible potential participants were provided with a consent form to be read and signed before commencement of the interview. Participants were informed that prior to recruitment, the study was granted ethics approval from the Office of Research Ethics at Simon Fraser University.

Eligibility criteria for study required participants to be: 1) 18 years or older; 2) a holder of a provincial medical card; 3) someone who successfully pursued a surgical procedure outside of Canada that was neither a transplantation nor a reproductive surgery (as these procedures often involve third parties and raise distinct ethical issues); and 4) someone who paid privately for the surgery sought abroad and for whom the procedure was not performed based on a referral from a Canadian physician.

Canadians who had travelled for transplantation or reproductive procedures were not included in this study due to the additional ethical concerns with these practices that are not covered in the information tool.

3.2 Data collection

Twenty-four semi-structured interviews were completed between October 2012 and December 2012. We conducted as many interviews as possible during this two-month period, after which we planned to cease data collection. The interviews were all conducted over the phone and recorded for future transcription. The shortest interview lasted just over 20 minutes, with the longest lasting nearly an hour and a half. The majority of interviews lasted approximately 45 minutes. The range in interview times is due to the semi-structured approach which provided some questions to guide the discussion and ensure meaningful feedback, but allowed for freedom for the participants to speak openly, which may broaden the understanding of the topic at hand [30].

All interviews were conducted by the first author using a semi-structured interview guide. The guide contained questions divided into four parts. The first part addressed basic background information of the participant including demographics, general health, and past travel experience. The second part asked questions relating to the participants' general decision-making behaviour. The third section asked questions related to participant's experiences as a medical tourist, particularly in terms of how they made the decisions relating to their trip. Both the second and third section probed the participant about the types of information they were considering during their decision-making process as a medical tourist. The final section of the interview guide contained questions asking for feedback on a draft version of the information sheet (see Figure 2), particularly the usefulness of such a document in the participants' context and recommendations for improving the document.

3.3 Data analysis

Interviews were transcribed verbatim. The first author reviewed all of the transcripts and extracted basic demographic information, basic procedure and trip information, as well as general feedback (positive or negative) on the information tool for each participant into a spreadsheet. She then assigned 6 transcripts for independent review by each team member. Three of these transcripts were the same across all members while three were unique. This strategy was used to minimize redundancy in transcript review and ensure that all members would encounter the range of opinions shared by participants. Following transcript review, we met to discuss emerging themes in participant feedback on the tool. Given the practical or applied nature of this formative study, the themes identified were primarily a priori or deductive in nature in that they focused on tangible improvements to specific aspects of the informational tool. After these themes were identified, the first author hand reviewed the transcripts and extracted segments that pertained to each into a Word document organized under six headings: positive feedback, negative feedback, specific recommendations, impacts of the tool on decision-making, potential places to access such an information tool, and type/tone of information conveyed in the tool. Verbal and written consent was obtained from interview participants for the use of anonymized quotes in published materials.

4. RESULTS

Overall, the 24 interview participants ranged in age from 24 to 65. Nine participants identified themselves as male and the remainder as female. Fifteen of the participants had traveled out of Canada for the Chronic Cerebrospinal Venous Insufficiency (CCSVI) procedure, also called renal or jugular angioplasty by participants. The remaining nine participants accessed a lumbar disc replacement surgery, a dental surgery, a gastric band surgery, a tattoo removal through surgical incision, a vertical sleeve gastrectomy, an eye-lid surgery, a meniscus surgery, and a cholecystectomy. Five of the 24 participants indicated that they had existing family connections in the destination where they underwent the medical procedures, with four of these five participants having emigrated to Canada from that country. When examining destination choice amongst CCSVI patients, seven out of fifteen participants had traveled to the United States, with the remainder traveling to Bulgaria, Poland, Mexico, India, Germany, and Egypt. Non-CCSVI patients traveled to Egypt, Germany, Spain, Philippines, Venezuela, Thailand, Hong Kong, Romania, Mexico and India for surgery. Overall, five participants had participated in multiple medical tourism trips, with four out of five repeating the CCSVI procedure in a different location, and one travelling for three separate medical procedures in three different locations.

Medical Tourism & Me

Thinking about going abroad for surgery or other medical care? This is referred to as medical tourism. Here's some important information for you to consider before deciding whether or not to become a medical tourist.

My Health

1. *Did you know* that not all countries have the same hospital quality control standards we use here in Canada?
Before you decide whether to go, contact the Canadian Blood Services Agency to ask if they are aware of any blood supply or storage concerns in the country you are thinking about visiting.

2. *Did you know* that your medical record here in Canada should include your entire health history, even for medical care given in other countries?
Before you decide whether to go, ask the hospital you are planning to book your procedure with if they will give you records in English or French to bring home with you.

3. *Did you know* that diseases contracted abroad could be brought back home with you?
Before you decide whether to go, visit a travel clinic and make sure you get all recommended vaccines.

My Home

1. *Did you know* that some types of medical care require follow-up appointments or aftercare treatment?
Before you decide whether to go, make sure that you will be able to get the aftercare or follow-up appointments you need by talking to your regular doctor here about this.

2. *Did you know* that before you consent to any medical care you should know the risks of the procedure?
Before you decide whether to go, be sure to obtain information on procedure risks, ideally from a trusted source.

3. *Did you know* that your own decision to go abroad for medical care could actually impact other Canadians as well?
Before you decide whether to go, consider that if you experience medical complications you may need to be treated for these in Canada and that doing so may lessen other patients' access to this medical care.

My Journey

1. *Did you know* that not all countries have the same patient and consumer protection laws we do here in Canada?
Before you decide whether to go, learn about malpractice laws in the countries you are considering visiting and make certain that your rights as a patient will be protected.

2. *Did you know* that some countries offer medical procedures to international patients that you cannot get here in Canada?
Before you decide whether to go, find out if the procedure you are considering has been shown to be safe and effective by a reputable agency.

3. *Did you know* that health care is a scarce resource around the world?
Before you decide whether to go, inform yourself about the most pressing health care challenges in the countries you are thinking about visiting and consider whether medical tourism may help or worsen them.

Be informed and share your thoughts, ideas, and experiences with others.
For more information go to: www.sfu.ca/medicaltourism/guide

Figure 2. Draft information sheet.

The 24 participants all read the information sheet prior to commencing the interview to allow for discussion on this document during the interview. The feedback provided by these participants is described in the remainder of this section. This feedback indicates the perceptions of these persons on the information sheet itself, with a particular focus on the format, usefulness and potential influence of the document on medical tourists' decision-making.

4.1 Feedback on format

Overall, almost all participants were generally positive when discussing the format of the tool. Regardless of the procedure sought or the country visited, participants had similar feedback on the format of the information tool, which was overwhelmingly positive. Positive feedback relating to format indicates that participants appreciated the concision of the document, including its short sentences and "*conversational*" language. Participants indicated that the document is "*easy to read*", relating this ease to both layout and language. Feedback on the format specifically indicates that the language is "*very clear*" and "*basic*". This was particularly important because many participants commented that the language and level of detail contained in the document must be easily accessible to persons with "*standard education*" and "*standard English speaking*" abilities. Furthermore, participants indicated that the short one-page format is favourable as the limited text is "*not boring*". The question-style format of the document was well-received by some participants who found this style encouraged persons to reflect on the information and take this information to do further research on their own, while others preferred that it be populated entirely with standard sentences.

While there were very few negative comments about the format of the document in terms of layout, participants provided formatting suggestions to improve the appeal of the document. Some participants recommended more of a checklist format over the question-style format. This format was recommended to avoid the document being "*wordy and plain*". Several participants also indicated that some of the repetition in wording was not appealing and made the document seem repetitive and condescending. The most common recommendation for improving the format of the document was to include pictures, images or more colours to increase the visual appeal.

In terms of language, while all participants agreed that the language was easy to understand, some participants pointed out areas where the information provided did not lend itself to specific actions. These participants demonstrated frustration with being unsure what the information in the tool was trying to convey. One participant indicated that education level would really impact whether a *"person's going to give it a second thought or not"* and suggested that while the language may be clear, the intended information may not be fully understood by all persons. Furthermore, some participants suggested that readers may have made up their mind before accessing the tool and may not have an open mind when reading the document, indicating challenges in conveying intended information through the format of an information tool.

Another suggestion by participants for maintaining the concision of the document while providing useful information was the inclusion of small anecdotes or stories as part of the format would be useful. Participants indicated that describing personal experiences through anecdotes helps empathize with the reader by creating a *"personal touch"*. They felt that a format with anecdotes illustrating both positive and negative experiences would prompt persons to do further research while creating a neutral tone.

4.2 Feedback on usefulness

Positive feedback on the usefulness of the information in the tool was fairly congruent, with similar feedback coming from participants who had travelled to different countries for distinct procedures. In general, participants provided positive feedback on the usefulness of the document in prompting further research by those considering medical tourism. Many participants stressed the importance of medical tourists doing *"as much research as you can"* after reading the document and felt that this document is successful in *"giv[ing] you things to think about"* while engaging in this research. Some participants indicated specific examples of points of information from the document that they had not previously considered that encouraged serious contemplation when reading the tool. Participants that indicated the tool would be useful for promoting further contemplation and research of medical tourism generally felt that the information was relevant to anyone travelling out of country for medical care, regardless of procedure.

Only a small minority of participants indicated that they did not think the informational tool would be useful for Canadians considering medical tourism. There were two main bases for this perspective: 1) even if the document does prompt a reader to do further research, it does not provide enough guidance or insight into how to do this research, or 2) the information presented in the document is not relevant to their experience.

Regarding the first basis, some participants indicated frustration after reading this document and wanting to know more about the issues being raised but not being sure of where to turn to access it. While the tool does provide a link to our own research website at the bottom of the page, some participants would have liked to have seen links to sources of information for each point. Furthermore, some participants, particularly CCSVI patients, were unsure what to do if they wanted to go abroad for surgery that they were denied access to domestically in that they felt this circumstance raised unique issues. Overall, participants who were unsure as to how to do further research after reading the document felt that the tool was "*unrealistic*" as it lacked the sufficient amount of information necessary to make informed decisions regarding the points raised in the tool.

Participants who did not find the tool particularly relevant indicated that while they feel it is important to do a great deal of research before travelling as a medical tourist, often the motivations for travelling involve some level of desperation which really impacts upon the decision-making process. In other words, it was thought that this search for hope or push abroad out of desperation impacts the usefulness of the tool in that this reality far outweighs the importance of any of the prompts shared in the document. This desperation for medical treatment makes further research about the information contained in the document a "*bit of a tall order*". Many participants said that at the time of deciding whether or not to travel out of Canada for medical care, they were in a lot of pain and this would have impacted their abilities to spend time searching for additional information to guide them in their decision-making both in terms of energy levels and because it would be "*something else to worry about*".

Some participants did not agree with all of the potential negative impacts of medical tourism suggested in the tool. They were most critical of suggestions that local citizens may not benefit from medical tourism hospitals and clinics. These participants often cited witnessing locals being treated at the facilities they visited or the undeniable economic gain to the community as providing confidence that medical tourism does not negatively impact destination communities. One participant said that he or she was "*keeping the faith that us being there would provide more money*" into the local economy, indicating that she assumed an overall positive benefit of medical tourism on the local community. Likewise, some participants disagreed with points in the document related to potential negative impacts on Canada, indicating that leaving Canada for care would provide overall benefits such as "*reduce[d] waiting times*" and "*decrease[d] costs*".

4.3 Potential influence of the information tool on decision-making

When participants were asked whether or not the information tool would have impacted their decision-making, many indicated that they probably would have still travelled out of country for their medical procedure but that it would have prompted them into conducting more research before leaving Canada. They indicated that the document might specifically impact upon intended users' expectations by encouraging further research, the outcome of which would provide more realistic expectations when considering or planning a medical tourism trip. On the other hand, some participants clearly stated that they would not have considered the information in this tool, either because they perceived that the document was implicitly and explicitly biased against medical tourism that would have turned them off from reading the tool, or because they had already considered all of the information contained in this document.

When asked directly about ethical prompts included in the tool, most participants said they were not apparent to them in their review of the document. One participant said that it did not raise any ethical concerns to him or her because the tool *"is just stating a fact"* and *"not [presenting medical tourism] as either pro or con"*, indicating that the perceived neutrality of the document impacted her or her perception on whether the information presents ethical concerns. The few participants that felt that the information in the tool raised ethical concerns did not see this tool as being explicit enough when raising these concerns, with one stating that *"the document doesn't really bring out the ethical concern of monopolizing [health] resources"* and another saying *"I don't see how this document raises those type of concerns; however, they're [concerns about patients in destination countries accessing medical] very valid concerns... that could be an ethical issue. But, if you bring it back to the document, I don't see how that's really linked to the type of questioning you ask in the document"*.

5. DISCUSSION

The feedback provided by Canadians that had previously travelled for medical tourism demonstrate varied opinions regarding the format, usefulness, and potential impact of an information tool for Canadians considering engaging in medical tourism. Although there was not complete agreement amongst all participants in terms of their opinion of the tool, the findings shared above demonstrate recurring themes throughout the feedback. While feedback on tool format was overall quite positive, feedback on usefulness and impact of the tool was divided between a few different opinions that ranged from strongly positive to strongly negative. The positive feedback indicated that the concision of the tool and language accessibility were generally appealing to participants. Furthermore, the overall positive feedback on usefulness of the tool indicated that as a source of information that prompts further research and contemplation about medical tourism prior to coming to a final decision, this tool will be of interest to a wide range of persons considering medical tourism as it is not guiding any particular action. This positive feedback leaves us encouraged a broadly disseminated tool would be appropriate for the intended target audience and could result in increased awareness of ethical concerns of medical tourism.

Negative feedback about the tool identified potential barriers to increasing awareness of ethical concerns for medical tourism. First of all, strong negative feedback on the document focused on the perceived negative or biased tone of the document and indicated that this might dissuade people from considering the information presented in the document. However, we expect that participants may have been more likely to view the tone as negative given that they had already travelled as medical tourists and may be more likely to reject information indicating potential negative impacts of medical tourism or to feel defensive around the choice they made [12]. For example, many participants described in detail their experiences accessing high quality medical care as a medical tourist and this may have biased their view on the potential safety concerns related to this practice. Similarly, participants that disagreed with potential negative impacts on access to medical care for destination populations might have been impacted by their experiences with a particular community, which might not be the case for persons that have not already travelled as a medical tourist.

Negative feedback on the usefulness of the document did highlight a lack of 'know-how' for what to do next with the information provided in the tool. Even when participants acknowledged that this tool increased their awareness of potential impacts of medical tourism, many were unsure how this might translate to action. Participants experiencing chronic pain said that, given their desperation to access care, their focus on improving their individual quality of life would likely have prevented them from considering potential societal impacts had they read this document before travelling for care. These results are similar to previous studies that have found Canadian medical tourists are less inclined to consider potential societal impacts resulting from medical tourism when asked about ethical concerns, and more likely to discuss concerns about the Canadian health care system contributing to their decision to participate in medical tourism [14]. In fact, medical tourists are most likely to travel for medical care in response to their experiences of feeling 'abandoned' by their own health care system [31]. As the majority of participants in this study were diagnosed with multiple sclerosis and travelled outside of Canada to access an unavailable procedure, many participants' likely experienced high levels of frustration trying to access care in the Canadian health care system before deciding to travel outside of Canada for the CCSVI procedure. The interviews demonstrated that the suggestion to contemplate potential ethical and health equity impacts of medical tourism on the Canadian health care system or health care systems globally may further frustrate patients that feel abandoned by the health care system and are desperate to access care. These findings agree with previous research which describes medical tourists as escaping the constraints of a national health care system by expanding their health care system to a global scale [31].

Participants mentioned several times that their desperation resulted in quick decision-making around medical tourism, which may limit informed consent, an existing problem in medical tourism [2]. This tool intends to encourage further consideration of information regarding potential impacts of medical tourism but, as demonstrated in the results, this may be frustrating for participants who feel as though they are in a desperate situation, face time constraints, and/or are unsure where to begin this research. However, we believe this frustration, if voiced, could potentially have a positive impact by promoting a need for greater research and regulatory structures within the medical tourism industry that facilitate more ethical decision-making and informed consent by medical tourists. Research on policy development indicates that shaping the public debate may have a large impact on resulting policy outcomes [32] and an information tool such as this one can definitely play a role in shaping the public debate around medical tourism in Canada and increasing advocacy surrounding ethical medical tourism and health service provision and utilization.

Following analysis of feedback from interview participants, the research team utilized this feedback to further revise the information provided in the document. Furthermore, we took into consideration feedback on document format from the interview participants as well as consultation with existing one-page health communication documents provided to patients in travel health clinics to guide the design of a document for dissemination. This final document can be seen in Figure 2. Dissemination of this document will take place through distributing copies to travel health clinics, primary health care providers, health authorities, and other groups interested in providing this document to persons using their services.

5.1 Wider relevance

The results and discussion of feedback from interview participants provide insight into the challenges and useful considerations that researchers, policy makers, and health care providers may wish to contemplate when considering the communication of ethics in health care utilization and provision. Global health researchers have identified a need for the application of ethical frameworks and tools to public health to encourage more fair and just allocation of health resources globally [33,34]. Consumer guidelines utilize a format that has been effective at increasing public discussion about some of the existing ethical issues in practices such as tourism. This public discussion may serve to introduce and reinforce more ethical practices through increased awareness of one's social responsibilities [35]. This research study demonstrates that guidelines or information sheets tailored to health care users may provide an effective means of engaging persons accessing health care into conversations and further contemplation about one's social responsibilities when accessing health care. As such, this research may be of interest to persons interested in promoting more just and fair allocation of health resources.

6. LIMITATIONS

This study is limited in its ability to develop an understanding of persons' actions following the consideration of information provided in this document. As the methods utilized for this analysis involved one-time interviews with persons that had already travelled for medical tourism, these persons did not speak on any future actions that they might take. While this study explored contexts in which this information may not be considered by someone reading the document, for those that did consider this information useful to the decision-making process, this study is limited in its ability to provide insight into the actual use of this information in this decision-making process. As such, the ability for this study to understand the effectiveness of the information sheet is limited.

7. FUTURE DIRECTIONS

While the findings of this formative study suggest that an information sheet focused on ethical concerns with medical tourism can play a useful role in raising awareness about these concerns and advocating for more ethical practice in the industry regardless of procedure, more research is needed to further explore this potential given the limitations of this study. For example, feedback from persons that have not already travelled as medical tourists could provide additional insight on the potential impact of this information tool. Additionally, there was limited feedback provided by participants on their understanding or potential engagement with the information provided in the tool. While participants easily engaged in conversation about potential impacts of medical tourism in previous questions, when asked about their thoughts on whether the information in this tool presented them with any ethical concerns, the use of the word "ethics" seemed to limit participants' engagement. This may be due to participants feeling uncomfortable discussing the ethics of an activity in which they have already engaged. This more limited discussion of ethics resulted in some lack of clarity regarding comprehension of the information presented in the tool. This suggests that future research with persons contemplating medical tourism should focus on greater in depth-discussion about the impacts of such a tool in terms of engagement with the information presented within. Furthermore, an additional resource such as a website could provide greater details, including anecdotes and alternative methods of describing the ethical concerns of medical tourism. The link to the website could be provided on the information sheet, and further research could explore feedback on the usefulness of such a website.

8. CONCLUSIONS

This paper has explored the use of an information tool as a means of increasing individuals' awareness of ethical concerns in medical tourism to enable Canadians to make more informed decisions about private health care utilization abroad. With existing sources of information for medical tourists commonly demonstrating a lack of neutral guidance [12, 16], this information tool responds to a pressing, practical knowledge gap. According to feedback from interviews with former Canadian medical tourists, the tool we have developed has the potential to raise awareness of ethical concerns during the decision-making process. However, it is unknown at this point whether this awareness will lead to shifts in attitude and changes in behaviour at both the individual and societal level to contribute to social responsibility in health care provision and utilization, and ultimately improved global health equity [28]. Moving beyond the formative research presented herein, implementation of the tool and evaluation of its uptake and utilization will shed light on whether or not such awareness will lead to these types of shifts and changes.

ACKNOWLEDGEMENTS

The development of this paper was supported through a Catalyst: Ethics Grant from the Canadian Institute of Health Research (funding reference number 114797). VAC is also supported by a Scholar Award from the Michael Smith Foundation for Health Research.

AUTHORS' CONTRIBUTIONS

KA conducted 23 of the interviews while JS conducted 1 interview. KA reviewed all interview transcripts and the remaining authors reviewed 6 transcripts each. All authors contributed to the development of the interview guide, discussions of emerging findings from the interviews, and revisions to the information sheet. KA wrote the manuscript with input from all other authors. JS and VAC reviewed and approved the final manuscript. All authors read and approved the final manuscript.

REFERENCES

1. Johnston R, Crooks VA, Snyder J, Kingsbury P: What is known about the effects of medical tourism in destination and departure countries?: A scoping review. International Journal for Equity in Health. 2010, 9: 24-10.

2. Penney K, Snyder J, Crooks VA, Johnston R: Risk communication and informed consent in the medical tourism industry: a thematic content analysis of Canadian broker websites. BMC Med Ethics. 2011 12-17

3. Snyder J, Crooks VA, Johnston R, Kingsbury P: Beyond sun, sand & stitches: assigning responsibility for the harms of medical tourism. Bioethics. 2012, 27: 233-242

4. Lunt N, Carrera P: Systematic review of web sites for prospective medical tourists. Tourism Rev. 2011, 66: 57-67.

5. Advisory Committee on Genetic Modification: *Annual Report*. London; 1999,

6. Crooks VA, Turner L, Cohen G, Bristeir J, Snyder J, Casey V, Whitmore R: Ethical and legal implications of the risks of medical tourism for patients: a qualitative study of Canadian health and safety representatives' perspectives. BMJ Open. 2013

7. Connell J: Contemporary medical tourism: conceptualisation, culture and commodification. Tour Manage. 2013, 34: 1-13.

8. Hopkins L, Labonte R, Runnels V, Packer C: Medical tourism today: what is the state of existing knowledge?. J Public Health Policy. 2010, 31: 185-198.

9. Pocock NS, Hong Phua K: Medical tourism and policy implications for health systems: a conceptual framework from a comparative study of Thailand, Singapore and Malaysia. Globalization and Health. 2011, 7

10. Horowitz MD, Rosensweig JA, Hones CA: Medical tourism: globalization of the healthcare marketplace. Medsc Gen Med. 2007, 9: 33

11. Widdows H: Localized past, globalized future: towards an effective bioethical framework using examples from population genetics and medical tourism. Bioethics. 2011, 25: 83-91.

12. Ormond M: Shifting subject of health-care: place 'medical tourism' in the context of Malaysian domestic health care reform. Asia Pac Viewp. 2011, 52: 247-259.

13. Johnston R, Crooks VA, Snyder J: "I didn't even know what I was looking for...": a qualitative study of the decision-making processes of Canadian medical tourists. Asia Pac Viewp. 2012, 8: 23-

14. Turner L: 'First world care at third world prices': globalization, bioethics and medical tourism. BioSocieties. 2007, 2: 303-325.

15. Snyder J, Crooks VA, Johnston R: Perceptions of the ethics of medical tourism: comparing patient and academic perspectives. 2012, 1-9. Ethics: Public Health

16. Snyder J, Crooks VA: New ethical perspectives on medical tourism in the developing world. Dev World Bioeth. 2012.

17. Ormond M, Sothern M: You, too, can be an international medical traveler: reading medical travel guidebooks. Health Place. 2012, 18: 935-941. 10.1016/j.healthplace.2012.06.018

18. Crooks VA, Bristeir J, Turner L, Snyder J, Casey V, Johnston R: Understanding the Health and Safety Risks for British Columbia's Outbound Medical Tourists. 2011, Department of Geography: Simon Fraser University

19. Terry N: Under-regulated healthcare phenomena in a flat world: medical tourism and outsourcing. New Engl Law Rev. 2007, 29: 415-463.

20. Walley J, Lawn JE, Tinker A, de Fransisco A, Chopra M, Rudan I, Bhutta ZA, Black RE: Primary health care: making alma-Ata a reality. Lancet. 2008, 372: 13-19.

21. Ostlin P, Schrecker T, Sadana R, Bonnefoy J, Hertzman C, Kelly MP, Kjellstrom T, Labonte R, Lundberg O, Muntaner C, Popay J, Sen G, Vaghri Z: Priorities for research on equity and health: towards an equity-focused health research agenda. PLOS Medicine. 2011, 8: 11

22. Sen Gupta A: Medical tourism in India: winners and losers. Indian J Med Ethics. 2008, 5: 3-5.

23. Lunt N, Hardey M, Mannion R: Nip, tuck and click: medical tourism and the emergence of web-based health information. Open Med Informat J. 2010, 4: 1-11.

24. Chapman A: Health Care Reform: A Human Rights Approach. 1994, Washington, DC: Georgetown University press

25. Marmot M, Allen J, Bell R, Goldblatt R: Building the global movement for health equity: from Santiago to Rio and beyond. Lancet. 2012, 379: 181-188.

26. Von Schirnding Y: Health and sustainable development: can we rise to the challenge?. Lancet. 2002, 360: 581

27. Graces J, Rodenas F, Sanjose V: Towards a new welfare state: the social sustainability principle and health care strategies. Health Policy. 2003, 65: 201-215.

28. Parvanta C, Nelson DE, Parvanta S, Harner RN: Essentials of Public Health Communication. 2011, Jones & Bartlett Learning: Mississauga

29. Bernhardt JM: Communication at the core of effective public health. Am J Public Health. 2004, 94: 2051-2053.

30. Lefebvre RC, Flora JA: Social marketing and public health intervention. Health Educ Behav. 1988, 15: 299-315.

31. Denzin NK, Lincoln YS: The SAGE Handbook of Qualitative Research. 2011, Thousand Oaks, California: SAGE Publications

32. Perfetto R, Dholakia N: Exploring the cultural contradictions of medical tourism. Consumption Markets & Culture. 2010, 13: 399-417.

33. Waddell C, Lavis JN, Abelson J, Lomas J, Sheperd CA, Bird-Gayson T, Giacomini M, Offord DR: Research use in children's mental health policy in Canada: maintaining vigilance amid ambiguity. Soc Sci Med. 2005, 61: 1649-1657.

34. Farmer P: Pathologies of Power: Health, Human Rights, and the New War on the Poor. 2005, California: University of California Press

35. Maclean SJ, Brown SA: The social determinants of global health: confronting inequities. In Health for Some: the political economy of global health governance. Edited by Maclean SJ, Brown SA, Fourie P. 2009, Palgrave Macmillan: Basinstoke

36. Clarke N, Barnett C, Cloke P, Malpass A: Globalising the consumer: doing politics in an ethical register. Polit Geogr. 2007, 26: 231-249.

Chapter 3

What is known about the effects of medical tourism in destination and departure countries? A scoping review

Rory Johnston [1*], Valorie A Crooks[1], Jeremy Snyder[2], Paul Kingsbury[1]

[1]Department of Geography, Simon Fraser University, 8888 University Drive, Burnaby, British Columbia, Canada.

[2]Faculty of Health Sciences, Simon Fraser University, 8888 University Drive, Burnaby, British Columbia, Canada.

ABSTRACT

Background

Medical tourism involves patients intentionally leaving their home country to access non-emergency health care services abroad. Growth in the popularity of this practice has resulted in a significant amount of attention being given to it from researchers, policy-makers, and the media. Yet, there has been little effort to systematically synthesize what is known about the effects of this phenomenon. This article presents the findings of a scoping review examining what is known about the effects of medical tourism in destination and departure countries.

Methods

Drawing on academic articles, grey literature, and media sources extracted from18 databases, we follow a widely used scoping review protocol to synthesize what is known about the effects of medical tourism in destination and departure countries. The review design has three main stages: (1) identifying the question and relevant literature; (2) selecting the literature; and (3) charting, collating, and summarizing the data.

Results

The large majority of the 203 sources accepted into the review offer a perspective of medical tourism from the Global North, focusing on the flow of patients from high income nations to lower and middle income countries. This greatly shapes any discussion of the effects of medical tourism on destination and departure countries. Five interrelated themes that characterize existing discussion of the effects of this practice were extracted from the reviewed sources. These themes frame medical tourism as a: (1) user of public resources; (2) solution to health system problems; (3) revenue generating industry; (4) standard of care; and (5) source of inequity. It is observed that what is currently known about the effects of medical tourism is minimal, unreliable, geographically restricted and mostly based on speculation.

Conclusions

Given its positive and negative effects on the health care systems of departure and destination countries, medical tourism is a highly significant and contested phenomenon. This is especially true given its potential to serve as a powerful force for the inequitable delivery of health care services globally. It is recommended that empirical evidence and other data associated with medical tourism be subjected to clear and coherent definitions, including reports focused on the flows of medical tourists and surgery success rates. Additional primary research on the effects of medical tourism is needed if the industry is to develop in a manner that is beneficial to citizens of both departure and destination countries.

1. BACKGROUND

Medical tourism involves patients leaving their country of residence outside of established cross-border care arrangements with the intent of accessing medical care, often surgery, abroad [1]. Medical tourists cross national borders to access care, motivated by health service issues such as high costs, lengthy wait times and/or a lack of accessibility in their home systems [2–4]. Popular accounts of medical tourism have emphasized its accessibility and novelty, evoking depictions of recovering from bargain-priced surgeries on beaches [5–7], as shown in Figure 1. For example, a Canadian who had gone to Costa Rica for multiple cosmetic surgeries as a medical tourist said that "It was that best experience that anybody could have because while you are healing, you are in the sunshine...it is absolutely incredible" (n.p.) [8]. A medical tourist from the United Kingdom (UK) who went to South Africa for cosmetic surgery explained that "I went back to the guest house and rested [after surgery]. I didn't have to lift a finger. It was bliss" (p.38) [9]. Accounts such as these contribute to the 'exoticism' of medical tourism, and in many ways shift the focus away from the seriousness of having a surgical intervention while abroad.

Hospitals positioning themselves as leading medical tourism destinations are largely found within lower and middle income countries (LMICs), where favourable exchange rates have given them a competitive advantage in attracting price-conscious international patients [10, 11]. This development departs from an earlier pattern of trade in medical services, in which private facilities in high-income countries marketed expensive health services to the global elite [12, 13]. Furthermore, trade in medical tourism currently takes place outside of any regulatory framework. This has led some to view it as an expression of profit-driven, commodified private care, and has invited criticisms that it poses as a threat to the equitable delivery of health care worldwide [14–16]. These criticisms are supported by a recent World Health Organization report on health equity that calls for universally available, primary care-oriented health care systems [17]. Specifically, medical tourism requires significant investment in private, for-profit tertiary care infrastructure.

Medical tourism hospitals in LMICs have worked hard to shed the negative images and stereotypes associated with the 'Third World' that are often held by residents of high income countries [18, 19]. This includes shifting the perception that goods and services produced in LMICs are of inferior quality [20–23]. The rise of the internet has been integral in this shift, as it has allowed destination hospitals to distribute details about facilities, staff and services to a previously unimaginably wide audience, while maintaining tight control over the presentation of information. For example, websites prominently feature the credentials of physicians, many of whom have trained in prestigious institutions in high income nations [24, 25]. Also featured are international safety accreditations that hospitals have secured to further build the confidence of prospective clients [25–27]. Among other things, the promotion of medical tourism via the internet and other mediums has aimed to overcome the reservations international patients might have regarding going abroad for surgery. Meanwhile, numerous risks are inherent in the practice of medical tourism. The possibility of developing embolisms due to flying long periods, especially post-operatively [28], disruptions to patients' continuity of care [29] and the lack of adequate malpractice laws in many destination countries are all risky prospects for medical tourists [30]. These risks have not stopped the steady growth of the industry [31, 32].

Coinciding with the growth of the medical tourism industry has been the attention its expansion has received from researchers, policy-makers, and the media alike. Yet, there has been little effort to systematically synthesize what is known about the effects of this practice, even though such a synthesis has the potential to inform research and policy agendas alike. To address this significant knowledge gap, in the remainder of this article we present the findings of a scoping review that sought to answer the question: what is known about the effects of medical tourism in destination and departure countries? Destination countries are conceptualized as those that medical tourists go to for procedures, while departure countries are their home nations. Herein we understand medical tourism to be the intentional pursuit of non-emergency medical treatment outside of a patient's home country that is likely to include a pre or post-operative stay abroad during which some tourist activities may be undertaken. It falls under the larger rubric of 'health tourism' and has a number of distinct streams, demarcated by procedure type. Popular examples are orthopaedic, dental, cardiac, cosmetic, reproductive and transplant tourism [29, 33–35]. Formal cross-border care arrangements between countries are excluded from this understanding of medical tourism, as such out-of-country care does not capture the patient-initiated nature of the practice. Also excluded are vacationers seeking emergency care while overseas, ex-patriates receiving treatments in their country of residence and spa vacationers. Distinctions of this nature have been made elsewhere [16].

Figure 1. The tourist element of medical tourism. Photo credit: authors. This photo shows the beachside view from a high-end resort in Mahabalipuram, India that is affiliated with a hospital treating international patients in a nearby city. Patients can choose to go to this resort during their recovery stay. The photo depicts the popularized image of medical tourism, whereby patients recover in desirable tourist locations after obtaining surgery in LMICs.

2. METHODS

In this article the findings of a scoping review are presented that address the question: what is known about the effects of medical tourism in destination and departure countries? Academic articles, grey literature, and media sources were gathered in a comprehensive fashion in order to answer this question. This is consistent with the scoping review approach, which aims to synthesize what is known about a particular issue across multiple literature types in order to achieve clarity about the state of knowledge and evidence that exists [36]. The synthesis presented here follows the scoping review protocol outlined by Arksey and O'Malley [37]: (1) identifying the question; (2) identifying relevant literature; (3) selecting the literature; (4) charting the data; and (5) collating, summarizing and reporting the results. In the remainder of this section the steps undertaken to complete each stage are outlined.

2.1 Identifying the Question and Relevant Literature

The first step of the scoping review was to hold a team meeting to identify a question that would serve as a useful contribution to scholarship on medical tourism. Following this meeting a search strategy was developed that was tailored to addressing the review question. Team members scanned the published literature on medical tourism to identify keywords relevant to the scoping question. Shown in Table 1, selected keywords probed five categories: (1) focus (e.g., medical tourism); (2) what (e.g., hospital); (3) who (e.g., patient); (4) why (e.g., effects); and (5) where (e.g., destination). Eight types of rationale were identified for the why category (addressing issues such as why the medical tourism industry operates and why patients decide on medical tourism), and keywords were generated for each. Countries known to be departure and destination points for medical tourists at the outset of the review were used to populate the where category.

After the keywords were finalized a database search strategy was developed in consultation with a librarian. The strategy was designed to scope the English-language grey, media, and academic literatures in order to find the breadth of sources that would assist with answering the review question. It was decided that combinations of keyword terms would be searched in 18 databases, summarized in Table 2. Academic and media databases required different search strategies. For academic databases, keywords across the five categories were searched using Boolean operators in order to maximize the combinations of terms ultimately scoped (shown as * in Table 1). It was found that some combinations of keywords generated unmanageable results (e.g., generating hundreds of thousands of sources). When such keyword combinations were identified the results were narrowed by the search manager through eliminating the term that had the broadest results. This strategy was found to successfully enhance the relevance of the identified sources.

Table 1. Scoping Review Keyword Search Strategy

Focus	What	Who	Why	Where
Medical tourism	Surger*	Patient	Decision making	Destination
Health tourism	Elective surger*	Tourist	Factors	Brazil
	Surgical Procedure*		Decision	India
	Hospital*		Attitudes	Thailand
	Clinic*		Motivation	South Africa
			Destination choice	Indonesia
			Tour*	Cuba
			Travel	Mexico
			Vacation*	Philippines
			Adventur*	Singapore
			Wait time	United States
			Wait list	Canada
			Queue	
			Speed	
			Value*	
			Ethic*	
			Privat*	
			Effects	
			Two tier	
			Cost savings	
			Affordability	
			Savings	
			Cost	
			Motivat*	
			Perspective*	
			Distance	
			Quality	

Table 2. Databases Searched for Scoping Review

Database Type	Database	Temporal Period Covered
Academic	Academic Search Premier	1984 - 20/08/2009
	Ageline	1978 - 10/08/2009
	Biomed Central	no recorded start date - 10/08/2009
	Business Source Complete	no recorded start date - 20/08/2009
	Canada Research Index	1982 - 21/07/2009
	CINAHL	1982 - 20/07/2009
	CPI.Q	1988 - 22/10/2009
	EconLit	1969 - 9/08/2009
	Geobase	1980 - 9/08/2009
	Global Health	1973 - 10/08/2009
	Medline	1950 - 9/08/2009
	PAIS International	1972 - 9/08/2009
	PsycINFO	1887 - 10/08/2009
	Sociological Abstracts	1963 - 20/08/2009
	Web of Science	1900 - 10/08/2009
Media	Alternative Press Index	1991 - 20/07/2009
	CBCA Current Events	1982 - 21/07/2009
	Canadian Newstand	1985 - 22/10/2009
	Lexis Nexis	no recorded start date- 22/10/2009

As noted above, the media database search strategy differed from that created for the academic databases. The team first had to identify the geographic scope of the media search. It was agreed that the best strategy would be to search for results pertaining to one country. This strategy was adopted as it was quickly realized that searching all media sources across the globe was not feasible, nor would it be necessary for the purposes of this scoping review. Canada was selected as the country of focus as this is where the research team is based, thus allowing the best access to Canada-focused media databases. Though the media sources included in the review are focused on Canada, they may replicate the types of discussions of medical tourism happening in other known high-income departure nations. 'Health tourism' and 'medical tourism' were the only terms searched in the media databases listed in Table 2. In addition, specific North American sources known to cover Canadian health services stories were searched within the Lexis Nexis database (namely the Globe and Mail, New York Times, Associated Press, Time inc., Washington Post, magazines, Toronto Sun, Toronto Star, CBC News). Across the searched academic and media databases, sources of all types (e.g., business briefs, research articles, newspaper editorials, industry reports) deemed relevant were retrieved. Retrieved sources were stored using the Refworks bibliographic management program.

2.2 Selecting the Literature

Titles and abstracts of sources identified by the search strategy were independently reviewed by team members in batches in order to select literature for inclusion in the scoping review. After completion of the independent review, team members met and sought consensus regarding whether a particular source was to be read in full. Consistent with the scoping review process, *post hoc* inclusion criteria were developed at this point [37]. Three bases for exclusion were identified, namely that there was either: (1) no focus on medical intervention, such as sources that dealt with health tourism more broadly (e.g., international travel to healing spas); (2) an exclusive focus on 'reproductive tourism' or 'transplant tourism', as the medical intervention (if any) in such cases is not restricted to the international patient and thus raises separate considerations; and/or (3) an overly general focus on international trade in health services or cross-border care, where there was no explicit reference to medical tourism. Any sources not in English were also excluded at this point. In instances where there was disagreement between team members as to whether or not a source should be reviewed in full, the source was discussed until consensus was reached. Such disagreement was minimal as the level of agreement among team members was high from the outset.

Once the title and abstract review was complete, included sources were reviewed in full. It should be noted that abstracts for media sources were not available and so all were reviewed in full. The reference lists of sources accepted for full review were hand searched and relevant sources not identified by the initial strategy that were deemed relevant to the review question were also reviewed in full. The exclusion criteria developed at the title and abstract review stage were again applied during full review. One additional criterion was added: if no 'data' (i.e., information points that contributed to answering the scoping question) were extracted from a source it was excluded. Every source identified for full review was read by two team members who were assigned by the search manager. Sources were reviewed in batches. As each batch was completed, the team met to make decisions regarding the inclusion or exclusion of sources. As with the title and abstract review stage, the level of agreement was high and any disagreement was resolved through seeking consensus among the entire team after discussion.

2.3 Charting, Collating, and Summarizing the Data

A spreadsheet securely hosted online and used by all team members was created to chart the data extracted from sources reviewed in full. Details regarding study design and sample (if relevant), publication information, and data pertinent to the scoping question were recorded in the spreadsheet. A series of meetings were held in order to discuss the extracted data and gain an overall perspective on the issues emerging from the literature. After this the lead author reviewed the extracted data and determined the overall themes that best characterized what had been gleaned from the sources reviewed. The identification of themes in the reviewed literature deemed relevant to the scoping question is an essential part of the charting process [37]. Team members reviewed the themes to confirm the interpretation that had been generated. The extracted data in the spreadsheet was next colour coded according to theme. This was done to assist with organizing how the findings of the scoping review would be reported. The final step was for team members to work together in order to identify key knowledge gaps emerging from the reviewed sources that have direct relevance to the scoping review question.

3. RESULTS

As shown in Figure 2, 291 sources were extracted for review from the 18 databases searched. A further 57 were found from hand-searching reference lists. Following the partial (title and abstract) and full reviews of these 348 sources, 203 were ultimately accepted for inclusion in this scoping review (a complete list of the sources included can be obtained by contacting the lead author). When compared with other methods of knowledge synthesis, such as systematic reviews, this is a very large sample with which to work. However, given that the goal of a scoping review is to characterize the breadth of knowledge surrounding a fairly open question [37], the size of the sample offers confidence that the review comprehensively and exhaustively captures what is known about the effects of medical tourism in destination and departure countries.

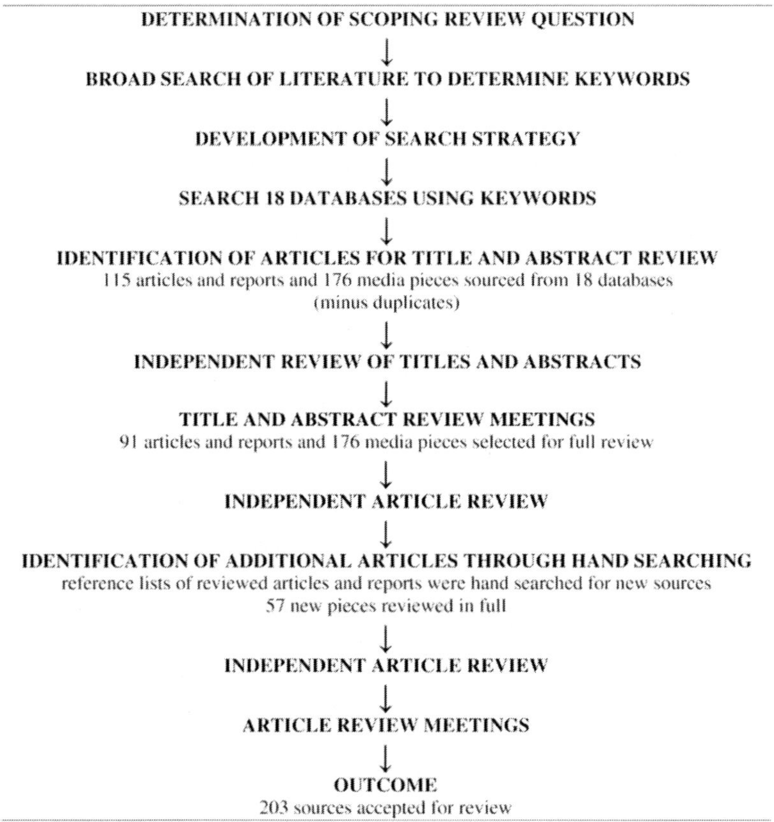

Figure 2. Scoping Review Search Strategy and Results.

Of the 203 sources included in the review, almost all were speculative in nature. A very small number (*n* = 6) were empirical studies reporting on primary data, while the majority (*n* = 107) of those included were from popular media sources. The remaining 91 are a mixture of reports, legal reviews, commentaries, editorials and business briefs. Most sources offered a perspective on medical tourism from the Global North, especially from the United States, focusing specifically on the flow of patients from developed nations to LMICs in the Global South. This significantly shaped discussion of the effects of medical tourism on destination and departure countries. Details of these effects, as gleaned from the reviewed sources, are broadly characterized by five crosscutting themes. These themes frame medical tourism as a: (1) user of public resources; (2) solution to problems; (3) revenue generating industry; (4) standard of care; and (5) source of inequity. As can be seen from these themes, the focus of discussion is on system-based effects rather than individual-level ones. In the remainder of this section we elaborate on these themes, exploring their positive and negative effects in both destination and departure countries.

3.1 Medical Tourism as a User of Public Resources

In discussions of the relationship between medical tourism and its influence on the allocation of public resources in destination countries, the practice has regularly been treated as a threat to the equitable use of public funds. Most commonly discussed has been India's rapidly growing private medical tourism industry, which has been aggressively cultivated with the aid of government funds [27, 38–42]. Government subsidies have been provided in the form of the provision of public land, corporate tax breaks and reduced tariffs on imported medical equipment for use in private hospitals serving medical tourists [38, 39, 42]. Though these private hospitals are required to provide some health services for impoverished domestic patients, there have been reports that these regulations are token requirements that regularly go unmet [43]. The use of public resources for medical tourism also extends to nations' publicly-financed health care services, with countries such as Singapore and Cuba opening public hospitals to privately-financed foreign patients [21, 44–47]. A less direct form of public subsidization, the hiring of physicians trained in public education systems by private medical tourism facilities is another example of a potentially inequitable use of public resources [38, 48]. Furthermore, physicians in LMICs who might normally practice in resource-poor environments can instead treat high-paying international patients, thereby gaining access to advanced technologies and superior facilities while receiving a higher wage [38, 49]. There has been some direct rebuttal of this concern by those who note that many doctors working at medical tourism facilities in LMICs were educated in high-income countries and are, in fact, returning to practice in systems they had previously left behind [11, 50].

Medical tourists may face major medical complications upon return to their home countries or require some form of post-operative care. A number of sources reviewed raised the issue of this care being a diversion of public resources, particularly when a medical tourist's domestic system is partially or fully funded through public financing [1,51]. For example, an Australian study demonstrated that treating an acutely infected total joint arthroplasty that had been performed abroad resulted in costs that were nine times what would have been spent had the patient received the procedure in a domestic hospital [52]. While this case vividly depicts the possible public costs of post-operative care stemming from medical tourism, the authors note that even with a 5% failure rate, the cost savings potential offered by medical tourism are so substantial that the Australian health care system would save public money by allowing patients to pursue the operation overseas and receive reimbursement [52]. Patients in other countries with publicly-funded health care systems may also be reimbursed by their governments for pursuing some elective surgeries abroad, likely lowering cost barriers to engaging in medical tourism. For example, a very small number of Canadian medical tourists have been reimbursed by their provincial public Medicare systems upon returning home if, prior to departure, they were able to demonstrate that the care being sought was medically necessary and unable to be provided in a timely manner in Canada [31, 53, 54]. However, in such cases medical tourism becomes a viable option only for those who can afford to first pay out-of-pocket.

3.2 Medical Tourism as a Solution to Health System Problems

Development of the medical tourism industry in LMICs carries with it a number of potential benefits that work to address some existing health system problems related to infrastructure development and retention of health human resources. Foremost amongst these is its purported capacity to spur both local and foreign investment into health care infrastructure [21, 41, 42, 55–57]. Use of such infrastructure may not be limited to medical tourists, thus benefiting local patients [19, 48, 58, 59]. In fact, most hospitals providing services to international patients are primarily dependent on locals for the bulk of their business [10, 11]. Investment into more advanced medical services in LMICs can also encourage patients who would otherwise travel abroad for care to get care at home, thus keeping capital within the country [58]. Another direct benefit resulting from the new infrastructure necessary for attracting medical tourists is the creation of long-term, high skilled jobs necessary for a strong tertiary health care system [1, 60]. It has also been suggested that the financial (e.g., high salaries) and technical (e.g., high technology work environments) incentives for doctors practicing in medical tourism can slow, or even reverse, the migration of locally trained health human resources abroad [1, 11, 19, 61]. Another often cited point is that medical tourism draws hard currency into LMICs, particularly from higher income countries [1,5, 10, 14, 62].

Medical tourism is depicted as a potential solution to an array of problems facing health care systems in patients' home countries, which can be most briefly summarized as those of the cost and affordability of care, wait times and access to medical facilities and treatments. The most commonly discussed among these is the issue of unaffordable care. Medical tourism is often presented as providing an option to patients in countries where medical care is prohibitively expensive [63, 64]. An extension of this view is the potential for a vigorous medical tourism industry to drive down domestic costs and encourage price transparency by introducing competition into countries with captive private markets, such as the United States [63, 65]. In countries with publicly financed health care systems, most notably the UK and Canada, medical tourism has been discussed as a solution for reducing long surgical wait lists [3, 31, 55, 59, 66]. For citizens of LMICs, it can offer access to needed care abroad that is simply unavailable at home due to inadequate medical infrastructure [10, 67]. For these patients, receiving care in another country may be their only option, and contrary to the levity of the label 'medical *tourism*', it may prove to be an impoverishing experience [68, 69].

3.3 Medical Tourism as a Revenue Generating Industry

The medical tourism industry is a lucrative source of hard currency for destination countries. Its annual growth potential has regularly been estimated above 25% for LMICs such as India, Malaysia and Thailand [2, 18, 70]. However, estimates of patient and currency inflows and their potential for growth are wildly varied, as definitional issues confound the already difficult attempt to determine the scale of the industry [2,3, 71, 72]. Examples of this include: estimates for the entire global medical tourism industry have been as low as 60,000 patients annually [2], while estimates of the annual American outflow alone have been as varied as 50,000 to 500,000 [63]. Perhaps the best example of speculative reporting has been the widely cited Deloitte report, which estimated an American patient outflow in 2007 at 750,000 and projected its increase to rise to 10,000,000 by 2012 [19]. Regardless of the actual numbers, it is generally agreed that the industry is growing rapidly in many Asian and South American nations, and is poised for significant growth in countries with the necessary human and technical resources [1, 10, 58, 73, 74]. India and Thailand's introduction of expedited medical visas is a sign of deep commitment to facilitating trade in medical tourism [40, 48, 70]. Thailand has gone so far as to try and negotiate the portability of public insurance between countries in the region to enhance the flow of inbound international patients [75]. Medical tourists are particularly valuable because of the money they invest in the local economy when compared with traditional tourists. For example, it has been estimated that medical tourists visiting Asia spend over twice as much as traditional tourists [76]. Cuba is an example of a country that established itself as a destination early on in the development of medical tourism, and it has built its industry into a revenue generator that has been used to fund its public health care system [45]. There has been a similar instance of providing care at a premium for foreign patients and using the difference to subsidize care for locals at a non-profit hospital in India [77].

Accompanying revenue gains in destination countries have been associated revenue losses in medical tourists' home countries. While the scattered estimates of international patient flows calls into question the reliability of estimates that find the United States currently losing billions of dollars annually to medical tourism [78, 79], it is only logical to assume that every patient leaving one health care system for another is taking their capital elsewhere. In response to the travel and care coordination needs of medical tourists, a cottage industry of brokerages has sprung up within patients' home countries. These brokerages coordinate the necessary travel, medical, accommodation and holiday arrangements for medical tourists, in exchange for fees from patients or a commission from the hospitals on a per-referral basis [3]. These brokerages have rapidly expanded to fill a niche and are often based in patients' home countries, thus further spreading the global reach of industry profits. This expansion has been facilitated by the spread of the internet, which has made medical tourism an increasingly accessible option and has helped Canada alone support more than 20 different brokerages [14].

3.4 Medical Tourism as a Standard of Care

A number of changes to standards of care have accompanied the development of medical tourism in destination countries. Perhaps the most noticeable has been the adoption of international accreditations that are modeled on Western standards [26,56, 80]. Foremost amongst these is hospital accreditation by the Joint Commission International (JCI), an offshoot of the primary body responsible for accrediting hospitals in the United States. The development and implementation of JCI has been credited with helping medical tourism to flourish by guaranteeing a standard of care comparable to that found in American hospitals in accredited hospitals [65]. This means of securing the trust of Western patients carries with it the potential for American standards to override local values and approaches to providing health care [15, 80]. However, the accreditation of an American-style quality of care does not extend to assurances of American-style liability. Most LMIC destination countries have limited malpractice laws and insurance requirements, leaving medical tourists with few opportunities for recourse should procedures go awry [1, 19, 27]. Another concern is the use of Western place aesthetics by medical tourism facilities, a phenomenon that has led to the displacement and homogenization of previously non-Western care spaces [15, 21]. This displacement includes designing hospital lobbies to look like shopping malls and patients' rooms to look like upscale hotels. Medical tourism hospitals are also able to deliver very high nurse-to-patient ratios, higher than those found in wards designated for local patients, due to the low costs of labour in LMICs [7,25, 35]. Medical tourists are also typically given the option of recovering at nearby resorts, offering amenities not replicated in the hospital environments that non-international patients may be required to stay in during recovery [1, 55, 81, 82].

Medical tourists returning to their home countries may bring with them altered conceptions of appropriate standards of care. Exposure to nurse-to-patient ratios that far exceed norms in most systems, hotel-like care spaces and the ability to choose one's physician from a website may result in patients demanding similar 'consumer as king' treatment at home [1, 83]. Expectations such as these are tied to the treatment of health care as a commodity, and may be especially deleterious for overburdened publicly-funded health care systems. More specifically, patients returning from abroad may wish to see elements of their privately-funded care instituted within their home systems [84, 85]. Medical tourism can also create a sense of limitless options for potential patients. If they find themselves unhappy with local options, the seemingly boundless variety of willing facilities and professionals available internationally holds with it the promise for effective treatment somewhere. This can encourage patients to treat medical tourism as a viable solution to medical problems, even if their prognosis is beyond treatment [68].

3.5 Medical Tourism as a Source of Inequity

Medical tourism has regularly been accused of exacerbating health care - and ultimately health - inequities in destination countries. One of these charges is that it exacerbates 'brain drain' within destination countries. The higher wages and advanced technolologies available at facilities offering medical tourism act as a lure for health care providers working in more modest facilities [40, 41, 48]. Because medical tourism facilities are primarily urban, this process also hastens the internal migration of health care providers from rural areas into cities, thereby enhancing rural deprivation [56, 75,86]. If the medical tourism industry achieves even a fraction of the flows of patients envisioned by early commentators, this could ultimately lead to locals being priced out of their own health care system, as demand from foreign patients can drive up the costs of providing care for everyone [14, 61, 66, 87]. The financial gains medical tourism offers LMICs has worked to shift health care investments into expensive, high-technology tertiary care that benefits a limited number of patients for the high cost outlay [39, 88]. This is especially inequitable given the large proportion of the population across LMIC destination countries for whom basic medical services are prohibitively expensive or simply unavailable [3, 40].

Medical tourists traveling from LMICs may end up exhausting their finances to access care, impoverishing themselves and their families, and thereby creating or exacerbating inequities [68, 69]. As long as the option for treatment exists elsewhere, departure countries lacking medical services may feel little pressure to address shortages, pushing citizens to risk material wellbeing in the pursuit of needed care [68,69]. When patients from countries with strong publicly-funded health care systems engage in medical tourism they undermine the ethos that has allowed these very systems to develop and survive [89]. They do this by effectively introducing a second tier of care available only to those with the ability to pay, an option that runs contrary to the provision of exclusively publicly-funded health care. Their actions extend further than the appearance of inequity, as each medical tourist puts his/her support behind the private provision of health care and a conception of health care as a commodity [1,39, 44]. This also applies to countries lacking in public health care, as the affordability of medical tourism works to neutralize pressure for reform and the equitable distribution of resources [90].

4. DISCUSSION

The results of this scoping review make it clear that what is currently known about the effects of medical tourism is minimal and that speculation abounds. Medical tourism is a phenomenon born out of a steadily globalizing economy, and is an expression of private, for-profit health care. The potential inequities this model of care can cultivate in both departure and destination countries can work to damage the provision of publicly-funded care where it is established and inhibit its development where it is not [14, 16,91, 92]. Conversely, the potential for medical tourism to operate in a mutually beneficial and equitable manner exists. This, however, requires oversight and regulation that guarantees meaningful compensation to the citizens of countries that offer their medical services to international patients [48, 59, 60].

What is known of medical tourism's effects is most effectively summarized by the five themes identified and reviewed above. These themes have particular implications for three global spheres of activity common to both departure and destination countries, namely those of their: health care systems, health and social policies and involvement in the medical tourism industry itself. These implications are expanded upon in this section, with consideration also being given to knowledge gaps identified by, and the limitations of, this scoping review.

4.1 Implications of the Effects of Medical Tourism

The sources reviewed indicate that medical tourism has, and will continue to have, significant effects on the health care systems of both departure and destination countries. For health administrators in patients' home countries, the lack of surveillance and monitoring of the practice of medical tourism can result in unaccounted for 'leakage' of patients outside of their jurisdictions [27, 32]. In systems that ration care, this negatively impacts the ability to accurately anticipate and distribute health care resources effectively. This could result in a situation where a health care system is perceived to be unresponsive to the needs of returning medical tourists due to inadequate resources being allotted for follow-up care. Medical tourism also works to introduce and normalize profit motives within the 'cultures' of the health care systems it engages with. For example, there are accounts of this approach to care resulting in the substitution of clinical factors with financial ones in treatment decision-making amongst medical tourists [52, 75]. This approach to decision-making may also negatively impact the local population in destination countries should a culture of compassion and necessity amongst health care professionals become supplanted by one of financial opportunism and pragmatism.

Countries hoping to capitalize on the ability of medical tourism to spur investment into their domestic health care infrastructures in the short term may find themselves becoming structurally dependent on foreign sources of investment and income well into the future. This dependency can lead to the entrenchment of inequitable modes of health care delivery [30, 93]. In countries that have signed over access to health care services under the international General Agreement in Trade in Services, this could bring about the permanent involvement of foreign parties with no vested interest in the well-being of local populations into their health care landscapes [14, 75]. Finally, the argument that destination countries should view investment into technology-intensive infrastructure for medical tourists as beneficial is undermined when the types of services being offered by hospitals are contrasted with the pressing health care needs of the local population. This contrast is all the more troubling when public money that could be used for wide-reaching, inexpensive primary health care initiatives is used to incentivize private investment into expensive tertiary care with a more limited impact.

Medical tourism has been likened to the 'wild west' with regard to the regulatory environment it occupies [12, 27]. This has resulted in what appears to be a lack of any discernable momentum to develop policy responses that can help to improve patient safety measures or guide the development of the industry in an equitable and ethical manner. As it stands, medical tourism is a largely characterized as an inequitable trade practice. LMIC destination countries offer essential services and skills that are needed for their own populations to foreigners with no vested interest in the success, failure or fairness of the systems they are using [39, 46]. There is, however, much room for improvement in such a young and unregulated industry. As an example, equitable buying guidelines could be established to assist patients and providers with better assessing the potential impacts of the medical tourism industry. Equitable buying guidelines have been developed and implemented in other global industries that have been implicated in inequitable business practices, and could be mirrored in the medical tourism industry [94, 95]. Likewise, patient safety could be greatly improved by developing stronger malpractice laws in destination countries and better informational and care coordination tools amongst departure countries. Implementing international standards for surveillance and monitoring of medical tourism would also allow for better care and planning by governments of both departure and destination nations.

Important questions are raised by this scoping review about the current organization of the medical tourism industry. The lack of regulation and standards amongst medical tourism brokerages have allowed for anyone to establish a brokerage [96–98]. For medical tourists who lack the confidence, knowledge or skills to arrange a medical tour on their own, private business owners who lack technical medical knowledge while having a financial stake in securing patients are, at the same time, their gatekeepers to care [3, 11, 14, 46]. While there is no evidence in the literature reviewed suggesting that there are problems with predatory or irresponsible business practices amongst medical tourism brokerages, regulatory voids around essential services set the stage for dangerous business practices that would be better pre-emptively managed.

Some medical tourism hospitals have ardently sought accreditation by trusted third party assessors such as JCI. This process is an indication that there is a willingness to self-regulate within the industry, and it raises the question of whether or not existing accreditation schemes should be altered or new schemes should be created that can assure consumers of equitable trade practices. Examples of these assurances could include meaningful levels of pro bono service for locals by medical tourism facilities, the creation and maintenance of primary health care clinics in under-served regions and other such measures to encourage health equity in destination countries.

4.2 Knowledge Gaps

The scoping review process demonstrated that there are major gaps in what is known about the effects of medical tourism. Most broadly, there is currently a very limited body of empirical research on the topic. In general, the reviewed sources revealed that*initial* estimates and ideas about medical tourism were heavily cited and recycled until they became treated as facts, both within and outside of scholarly publications. This suggests that existing sources must be approached carefully, lest misrepresentation be perpetuated further. The dearth of research-based knowledge about medical tourism does, however, provide a 'green light' for the development of new research into the phenomenon. The need to develop research is pressing given the number of potential impacts medical tourism poses for the equitable delivery of health care in departure and destination countries alike, along with the other effects this practice holds. In effect, research from across the social and health sciences is necessary to begin to properly document and meaningfully explore this trade practice. In the remainder of this sub-section a number of specific knowledge gaps are expanded on so as to identify possible 'ways ahead' for researchers interested in exploring medical tourism.

First and foremost, there is no reliable hard data on patient numbers, flows, treatment types and success rates. As it stands, there are conflicting estimates of patient numbers, with magnitudes of difference between them that likely stem from definitional issues of what comprises medical tourism. Success rates are reported by some facilities, but without an independent auditor or the inclusion of long-term success rates after medical tourists return to their home countries, these are of little value. This contrasts sharply with the glowing reports of treatment success rates in some business briefs, reports and popular news media that have failed to acknowledge the purely speculative and self-reported nature of the numbers at this point. Research is needed that can produce reliable understandings of the scope and volume of the practice of medical tourism in a way that is comparable across facilities and countries. Such research could also offer some of the first insights into important differences between medical tourists based on factors such as age, gender, citizenship, and health status, where such differentiation was simply not considered by the sources included in the scoping review. Destination countries that have introduced medical visas serve as a logical place to begin collecting this valuable data. The continuing failure of departure countries to introduce monitoring practices, especially using standardized units, makes it difficult to collect corresponding data at the other end of the medical tourism trail.

Legal and regulatory issues related to medical tourism must be expanded upon by researchers. As it stands, reviewed sources only managed to capture legal perspectives from the United States. While these sources likely offer a useful template for legal issues surrounding international trade between high income nations and LMICs, legal perspectives from other nations, especially those of destination countries, are needed to help to further clarify issues of malpractice and other legal dimensions of the industry. Regulatory frameworks are often cited as a necessity to ensure the safety of medical tourists, but sources that seek to realistically outline what these regulations would look like and how they would operate are absent. Meanwhile, these frameworks are necessary if the continuing expansion of the industry is to adopt safe and equitable approaches to organizing and delivering care.

Finally, issues of equity have largely been ignored in the popular, legal and business literature surrounding medical tourism, with reports instead focusing on the novelty of the industry or low prices available to potential patients. Within the academic literature, health equity has been more of a 'hit or miss' issue. For example, medical tourism has been consistently cited as a solution to both the problems of wait times and cost in departure countries by enhancing access to needed care, despite there being no evidence of it impacting either of these factors at the system level. In relation to this same point, much less attention has been given to understanding how such an approach to addressing system problems exacerbates inequities among residents of departure countries through enabling access to care abroad only for those who can afford it. Future research must make an effort to clarify what impacts medical tourism trade practices and financing arrangements are having on the delivery of health care within departure and destination countries, and at the very least provide the basic context for understanding how health care is currently arranged within them.

5. LIMITATIONS

Despite rigorous measures such as using multiple reviewers on all sources and seeking the guidance of a librarian in designing the search strategy, two major limitations are evident in the scoping review process employed. First, only English-language sources were considered for inclusion. Given the global nature of medical tourism, it is unlikely that all relevant literature is available in English. Second, media sources reviewed were restricted to those from Canada and a small number of major North American magazines and newspapers with relatively large Canadian audiences. This restriction was a necessary one, as the sheer volume of irrelevant popular literature to initially sort through in a non-restricted search would have rendered the scoping review unmanageable. Though focused on Canada, the media sources that were ultimately included provide a thorough illustration of the kinds of discussions about medical tourism being held locally, regionally and nationally in other high-income departure countries such as the United States and the UK.

6. CONCLUSIONS

This scoping review has identified five themes that effectively synthesize what is known about the effects of medical tourism in destination and departure countries. First, medical tourism can be understood as a user of public resources. This includes consuming public health care resources in destination countries through redirecting them to the private sector, and in departure countries through the provision of follow-up care for medical tourists. Second, the practice of medical tourism can be seen as offering solutions to problems. It has the potential to aid in the development of health care infrastructure in destination countries while reducing wait times and costs for care for residents of departure countries. Third, medical tourism is also a revenue generating industry. This industry can generate revenues for destination countries as a form of health services trade. Conversely, it can result in a net loss of capital to departure countries. Fourth, medical tourism can be seen as setting a standard of care. By seeking accreditation, destination countries may develop a Western-oriented standard of care, including in facility aesthetics. Due to low labour costs in destination countries, medical tourists may develop expectations of standards of health human resource provision that are unaffordable, and therefore unattainable, in high-income departure countries. The process of medical tourism can also contribute to the commodification of health care and a perception of the patient as consumer. Finally, the cumulative effects of medical tourism position it as a source of inequity. Within destination countries, it can contribute to an internal brain drain of trained medical workers from rural to urban areas and from the public to the private sector. Medical tourists can face a significant drain on their own financial resources and, by engaging in travel abroad for medical services, they may contribute to a loss of impetus for reform of their home health care systems.

The findings of this scoping review indicate that medical tourism is a very important phenomenon given its effects on the health care systems of departure and destination countries alike. Various commentators point to both the potential for negative and positive effects of medical tourism, marking it as a contested phenomenon. Without regulation, the potential positive effects may not be realized and medical tourism may contribute to a harmful global commodification of health care. These regulations must be developed, but the ability to do this is hampered by significant gaps in our knowledge of the effects of medical tourism, as identified by this review. To address these gaps, data regarding medical tourism must be subjected to clear and unified definitions of the phenomenon, flows of medical tourists and success rates of their surgeries must be clearly documented, and additional primary research into the impacts of medical tourism must be mounted. Until these research gaps are filled, medical tourism is likely to remain an increasingly visible and practiced global phenomenon, subject to much conjecture but little regulation or understanding.

ACKNOWLEDGEMENTS

Funding was provided by a Catalyst Grant from the Canadian Institutes of Health Research. Thanks also go to librarian Heather DeForest for consulting with the team on the search strategy.

AUTHORS' CONTRIBUTIONS

RJ led the identification of themes and writing of this manuscript. All authors contributed to the design of the scoping review, reviewing sources, and assisting with data management and interpretation. All authors provided feedback throughout the drafting of this article and have read and approved the final manuscript.

REFERENCES

1. Whittaker A: Pleasure and pain: Medical travel in Asia. Glob Public Health. 2008, 3:271.

2. Ehrbeck T, Guevara C, Mango PD, Cordina J, Singhal S: Health care and the consumer. McKinsey Quarterly. 2008, 80-81.

3. Leahy A: Medical tourism: The impact of travel to foreign countries for healthcare. Surgeon. 2008, 6: 260-261.

4. Horowitz MD, Marsek P, Mohanasundaram S, Pachisa M, Jones CA, Keith LG, Metaxotos N, Heng Boon Chin A, Yuen Tan Y, Yiun Teo S: Why in the world do patients travel for health care?. Asia Pac Biotech News. 2008, 12: 24-53.

5. MacQueen K: Ah, Cuba: sun, cigars and hip replacements. Maclean's. 2007, 120: 30

6. A nip and tuck with a safari on the side. Calgary Herald. 2008, F1-

7. Alsever J: Basking on the Beach, or Maybe on the Operating Table. NYT. 2006, 5.

8. Nip, tuck vacations risky, doctors warn. CBC News. 2007, [http://www.cbc.ca/consumer/story/2007/01/10/cosmetic-surgery.html]

9. Choat I: Surgically enhance your wage. Travel Weekly. 2003, 1688: 38-39.

10. Lautier M: Export of health services from developing countries: The case of Tunisia. Soc Sci Med. 2008, 67: 101-110.

11. Oberholzer-Gee F, Khanna T, Knoop C: Apollo Hospitals - First-World Health Care at Emerging Market Prices. 2007, Boston, Mass: Harvard Business School Cases

12. MacReady N: Developing countries court medical tourists. Lancet. 2007, 369: 1849-1850.

13. Hutson Gray H, Cartier Poland S: Medical Tourism: Crossing Borders to Access Health Care. Kennedy Inst Ethics J. 2008, 18: 193-201.

14. Turner L: 'First world health care at third world prices': globalization, bioethics and medical tourism. Biosocieties. 2007, 2: 303-325.

15. Saniotis A: Changing ethics in medical practice: a Thai perspective. Indian J Med Ethics. 2007, 4: 24-25.

16. Pachanee C, Wibulpolprasert S: Incoherent policies on universal coverage of health insurance and promotion of international trade in health services in Thailand. Health Policy Plan. 2006, 21: 310-318.

17. CSDH: Closing the gap in a generation: health equity through action on the social determinants of health. Final Report of the Commission on Social Determinants of Health. 2008, Geneva: WHO

18. Neelakantan S: India's Global Ambitions. Far East Econ Rev. 2003, 166: 52-54.

19. Operating Profit: Globalisation and Health Care. Economist. 2008, 388:

20. Mudur G: Hospitals in India woo foreign patients. BMJ. 2004, 328: 1338-1338.

21. Connell J: Tummy tucks and the Taj Mahal? Medical tourism and the globalization of health care. Tourism management: analysis, behaviour and strategy. Edited by: Woodside AG, Martin D. 2008, Wallingford, UK: CAB International, 232-244.

22. Tremonti AM: The Current. CBC Television. 2005, [Radio]

23. Loose C: Operation Vacation; Big Savings Have More Overseas Travelers Mixing Surgery With Sightseeing. Washington Post. 2007

24. Howze KS: Medical Tourism: Symptom or Cure. Ga L Rev. 2007, 41: 1013-1052.

25. Olian C: 60 minutes: Medical Tourism. CBS Television. 2006

26. Dunn P: Medical tourism takes flight. Hosp Health Networks. 2007, 81: 40-44.

27. Cortez N: Patient without Borders: The Emerging Global Market for Patients and the Evolution of Modern Health Care. Ind L J. 2008

28. Carabello L: A medical tourism primer for U.S. physicians. J Med Pract Manage. 2008, 23: 291-291.

29. Jesitus J: Safari surgery. Cosmetic Surg Times. 2006, 9: 1-14.

30. Burkett L: Medical tourism. Concerns, benefits, and the American legal perspective. J Leg Med. 2007, 28: 223-245.

31. Eggertson L: Wait-list weary Canadians seek treatment abroad. CMAJ. 2006, 174: 1247

32. Wolff: Passport to Cheaper Health Care. Good Housekeeping. 2007, 245: 190-190.

33. Turner L: Commercial organ transplantation in the Philippines. Camb Q Healthc Ethics. 2009, 18: 192-196.

34. Turner L: Cross-border dental care: 'dental tourism' and patient mobility. Br Dent J. 2008, 204: 553-554.

35. Cetron M, DeMicco FJ: Club Medic. Asia Pac Biotech News. 2006, 10: 527-531.

36. Davis K, Drey N, Gould D: What are scoping studies? A review of the nursing literature. Internat J Nurs Stud. 2009, 46: 1386-1400.

37. Arksey H, O'Malley L: Scoping studies: Towards a methodological framework. Internat J Soc Res Method. 2005, 8: 19-32.

38. Sen Gupta A: Medical tourism in India: winners and losers. Indian J Med Ethics. 2008, 5: 4-5.

39. Garud AD: Medical tourism and its impact on our healthcare. Natl Med J India. 2005, 18: 318-319.

40. Chinai R, Goswami R: Medical visas mark growth of Indian medical tourism. Bull WHO. 2007, 85: 164-165.

41. Chee HL: Medical Tourism in Malaysia: International Movement of Healthcare Consumers and the Commodification of Healthcare. 2007, Singapore: Asia Research Insitute

42. Mudur G: India plans to expand private sector in healthcare review. BMJ. 2003, 326: 520-520.

43. Sen Gupta A: Medical Tourism and Public Health. People's Dem. 2004, 28:

44. Ooi CS: State-Civil Society Relations and Tourism: Singaporeanizing Tourists, Touristifying Singapore. SOJOURN. 2005, 20: 249-272.

45. Charatan F: Foreigners flock to Cuba for medical care. BMJ. 2001, 322: 1198-1198.

46. Klaus M: Outsourcing Vital Operations: What If U.S. Health Care Costs Drive Patients Overseas for Surgery. Quinnipiac Health L J. 2006, 9: 219-247.

47. Yelaja P: Foreigners flocking to Singapore for state-of-the-art medical care. Toronto Star. 2007, AA1-

48. Arunanondchai J, Fink C: Trade in health services in the ASEAN region. Health Promot Int. 2006, 21: 59-66.

49. Kurlantzick J: Sometimes, Sightseeing is a Look at Your X-Rays. NYT. 2007, 10-

50. Badam RT: Americans, Europeans head to India for cheap, high-quality medical care. Guardian. 2005, B7-

51. Turner L: Medical tourism: Family medicine and international health-related travel. Can Fam Physician. 2007, 53: 1639-1641.

52. Cheung IK, Wilson A: Arthroplasty Tourism. Med J Austr. 2007, 187: 666-667.

53. Prashad S: The world is your hospital. Can Bus. 2008, 81: 62-66.

54. Medical tourism: need surgery, will travel. CBC News Online. 2004, [http://www.cbc.ca/news/background/healthcare/medicaltourism.html]

55. García-Altés A: The Development of Health Tourism Services. Ann Tour Res. 2005, 32: 262-266.

56. Connell J: Medical tourism: Sea, sun, sand and ... surgery. Tour Manag. 2006, 27: 1093-1100.

57. Shimo A: New hot spots for surgery. Maclean's. 2000, 120: 44-44.

58. Blesch G: Rising in the Persian Gulf. Mod Healthc. 2007, 37: 26-28.

59. Horowitz MD, Rosensweig JA, Jones CA: Medical Tourism: Globalization of the Healthcare Marketplace. Med Gen Med. 9: 33-33.

60. Importing Competition: globalisation and health. Economist. 2008, 388: 12-12.

61. Bajaj K: Come, Heal Thyself. Bus Today. 2007, 16: 139-144.

62. Hughes C: Cuba Lures Medical Tourists With Cheap Care, New Treatments. 1991, Associated Press

63. York D: Medical tourism: The trend toward outsourcing medical procedures to foreign countries. J Cont Educ Health Prof. 2008, 28: 99-102.

64. Save on surgery. Kiplinger's Personal Fin. 2009, 63: 18

65. Herrick DM: Medical Tourism: Global Competition in Health Care. 2007, Dallas, Texas: National Centre for Policy Analysis

66. Botten G, Grepperud S, Nerland S: Trading patients: Lessons from Scandinavia. Health Policy. 2004, 69: 317-327.

67. The Healing Touch. Economist Intel Unit. 2005, 13: 1-2.

68. Kangas B: Hope from Abroad in the International Medical Travel of Yemeni Patients. Anthropol Med. 2007, 14: 293-

69. Kangas B: Therapeutic itineraries in a global world: Yemenis and their search for biomedical treatment abroad. Med Anthropol. 2002, 21: 35-78.

70. Teh I: Healthcare Tourism in Thailand: Pain ahead?. Asia Pac Biotech News. 2007, 11: 493-497.

71. Hume LF, Demicco FJ: Bringing Hotels to Healthcare -- A Rx for Success. J Qual Assurance Hospitality Tour. 2007, 8: 75

72. Brown SB: Medical Tourism: Nations vie for health dollars. Hosp Health Networks. 2008, 82: 49-

73. McMurdy D: Vanity Takes a Vacation. National Post. 2007, FW2

74. Brennan R, Benzie R: Countries offer quick getaway from waiting lists. Toronto Star. 2004, H.04

75. Wibulpolprasert S, Pachanee C, Pitayarangsarit S, Hempisut P: International service trade and its implications for human resources for health: a case study of Thailand. Human Res Health. 2004, 2: 10

76. Jorgensen G: Beauty and the beach; Shorter wait times, lower costs are making Asia the hot spot for going under the knife. Toronto Star. 2007, T1-

77. Evans-Pritchard A: India poised for medical tour bonanza. Calgary Herald. 2006, A23

78. Medical Tourism retail clinics challenge status quo in US. PharmacoEconomics Outcomes News. 2008, 559: 15-

79. Crone RK: Flat Medicine? Exploring Trends in the Globalization of Health Care. Acad Med. 2008, 83: 117-121.

80. Jenner EA: Unsettled borders of care: medical tourism as a new dimension in America's health care crisis. Res Soc Health Care. 2008, 26: 235-149.

81. Marlowe J, Sullivan P: Medical Tourism: The Ultimate Outsourcing. Hum Res Plan. 2007, 30: 8-10.

82. O'Reagan S: Medical tourism: Reaction from health-care establishment. CTV-Canada AM. 2007, [Television]

83. City resident opts for hip surgery overseas. Burnaby Now. 2008, 5

84. Laidlaw S: Woman 'got my life back' after surgery in India. Toronto Star. 2008, L1-

85. Yelaja P: India offers surgery in a hurry. Toronto Star. 2006, A01-

86. Blouin C: Trade policy and health: from conflicting interests to policy coherence. Bull WHO. 2007, 85: 169-173.

87. Lee C: Just what the doctor ordered. Monash Bus Rev. 2007, 3: 10-12.

88. Ramírez de Arellano AB: Patients without borders: the emergence of medical tourism. Int J Health Serv. 2007, 37: 193-198.

89. Moore A: Health tourism. Don't forget your toothbrush. Health Serv J. 2009, 119: 18-20.

90. Unti JA: Medical and surgical tourism: the new world of health care globalization and what it means for the practicing surgeon. Bull Am Coll Surg. 2009, 94: 18-25.

91. Reed C: Medical Tourism. Med Clin North Am. 2008, 1433-1446. 10.1016/j.mcna.2008.08.001.

92. Rosies: Medical tourism: Traveller's dream or healthcare nightmare?. NIP. 2007, 4-4.

93. Carrera PM, Bridges JF: Globalization and healthcare: understanding health and medical tourism. Expert Rev Pharmacoecon Outcomes Res. 2006, 6: 447-454.

94. Moore G: The Fair Trade Movement: Parameters, Issues and Future Research. J Bus Eth. 2004, 53: 73-86.

95. Raynolds LT, Murray D, Leigh Taylor P: Fair trade coffee: building producer capacity via global networks. J Int Dev. 2004, 16: 1109-1121.

96. Forgione DA, Smith PC: Medical tourism and its impact on the US health care system. J Health Care Finance. 2007, 34: 27-35.

97. Pooley E: Have Ailment, Will Travel. Can Bus. 2006, 79: 68

98. Mirrer-Singer P: Medical Malpractice Overseas: The Legal Uncertainty surrounding Medical Tourism. Law Contemp Probs. 2007, 70: 211-232.

Chapter 4

Medical tourism and policy implications for health systems: a conceptual framework from a comparative study of Thailand, Singapore and Malaysia

Nicola S Pocock and Kai Hong Phua

Lee Kuan Yew School of Public Policy, National University of Singapore, 469C Bukit Timah Road, OTH Building, Singapore 259772, Singapore

ABSTRACT

Medical tourism is a growing phenomenon with policy implications for health systems, particularly of destination countries. Private actors and governments in Southeast Asia are promoting the medical tourist industry, but the potential impact on health systems, particularly in terms of equity in access and availability for local consumers, is unclear. This article presents a conceptual framework that outlines the policy implications of medical tourism's growth for health systems, drawing on the cases of Thailand, Singapore and Malaysia, three regional hubs for medical tourism, via an extensive review of academic and grey literature. Variables for further analysis of the potential impact of medical tourism on health systems are also identified. The framework can provide a basis for empirical, in country studies weighing the benefits and disadvantages of medical tourism for health systems. The policy implications described are of particular relevance for policymakers and industry practitioners in other Southeast Asian countries with similar health systems where governments have expressed interest in facilitating the growth of the medical tourist industry. This article calls for a universal definition of medical tourism and medical tourists to be enunciated, as well as concerted data collection efforts, to be undertaken prior to any meaningful empirical analysis of medical tourism's impact on health systems.

1. INTRODUCTION

Growing demand for health services is a global phenomenon, linked to economic development that generates rising incomes and education. Demographic change, especially population ageing and older people's requirements for more medical services, coupled with epidemiological change, i.e. rising incidence of chronic conditions, also fuel demand for more and better health services. Waiting times and/or the increasing cost of health services at home, coupled with the availability of cheaper alternatives in developing countries, has lead new healthcare consumers, or medical tourists, to seek treatment overseas [1]. The correspondent growth in the global health service sector reflects this demand. The globalisation of healthcare is marked by increasing international trade in health products and services, strikingly via cross border patient flows.

In Southeast Asia, the health sector is expanding rapidly, attributable to rapid growth of the private sector and notably, medical tourism, which is emerging as a lucrative business opportunity. Countries here are capitalising on their popularity as tourist destinations by combining high quality medical services at competitive prices with tourist packages. Some countries are establishing comparative advantages in service provision based on their health system's organizational structure (table 1). Thailand has established a niche for cosmetic surgery and sex change operations, whilst Singapore is attracting patients at the high end of the market for advanced treatments like cardiovascular, neurological surgery and stem cell therapy [2]. In Singapore, Malaysia and Thailand alone, an estimated 2 million medical travellers visited in 2006 - 7, earning these countries over US$ 3 billion in treatment costs (table 2).

Table 1. Health systems in comparison [3]

Country	Thailand	Malaysia	Singapore
Organizational structure	Pockets of excellence in some private Bangkok hospitals	Growing private health sector with movement of qualified workforce	Balanced public-private mix, corporatized public sector
National strategy	Regional health hub Extensive tourism infrastructure	Industrial strategy to develop tourism	Economic growth strategy to develop biomedical industries Regional service hub Medical R&D support
Policy impact	Issues of growing inequity and urban-rural divide	Public-private divide Racial inequities between public and private sectors	Narrow income gaps of public and private sectors

Table 2. Export of health services [2, 4, 5]

	Estimated earnings	No. foreign patients	Origin of patients (in order of volume)	Specialty
Thailand (2006)	Baht 36 billion (US$ 1.1 billion)	1.4 million	Japan, USA, South Asia, UK, Middle East, ASEAN countries	Cosmetic and sex change surgery
Singapore (2007)	S$ 1.7 billion (US$ 1.2 billion)	571 000	Indonesia, Malaysia, Middle East	Cardiac and neuro surgery, joint replacements, liver transplants
Malaysia (2007)	253.84 million MYR (US$78 million)	341 288	Indonesia, Singapore, Japan, India, Europe	Cardiac and cosmetic surgery

Carrera and Bridges (2006) define medical tourism as "the organized travel outside one's natural healthcare jurisdiction for the enhancement or restoration of the individual's health through medical intervention", using but not limited to invasive technology. The authors define medical tourism as a subset of health tourism, whose broader definition involves "the organized travel outside one's local environment for the maintenance, enhancement or restoration of the individual's wellbeing in mind and body". Importantly, their definition of medical tourism takes into account the territorially bounded nature of health systems, where access to healthcare is often but not always limited to national boundaries [6]. Medical tourism constitutes an individual solution to what is traditionally considered a public (government) concern, health for its citizens, who at the micro level are responding to market incentives by seeking lower cost and/or high quality care overseas that cannot be found at home. These tourists may be uninsured or underinsured. Travelling overseas for medical care has historical roots, previously limited to elites from developing countries to developed ones, when health care was inadequate or unavailable at home. Now however, the direction of medical travel is changing towards developing countries [7], and globalization and increasing acceptance of health services as a market commodity [8] have lead to a new trend; organized medical tourism for fee paying patients, regardless of citizenship, who shop for health services overseas using new information sources, new agents to connect them to providers, and inexpensive air travel to reach destination medical [9].The impact of medical tourism on health systems is as yet unknown due to a dearth of data and empirical analysis of the phenomenon.

Governments are noticeably playing a strong marketing and promotional role in the emerging medical tourism industry. This is a clear trend in Southeast Asia, especially in Thailand, Singapore and Malaysia, the main regional hubs for medical tourism, where medical tourist visas are available and government agencies have been established with the mandate to increase medical tourist inflows [10]. Governments in Indonesia, the Philippines and Vietnam have also expressed interest in promoting the industry. The potential economic benefits of medical tourism make it an attractive option for governments. Medical tourism can contribute to wider economic development, which is strongly correlated with improved population health status as a whole, e.g. increased life expectancy, reduced child mortality rates [11]. Encouraging foreign direct investment in healthcare infrastructure and medical tourist inflows with correspondent revenue can create additional resources for investment in health care [12].

Furthermore, medical tourism may slow or reverse the outmigration of health workers, particularly of specialists [13].

However, health systems in some of these countries face challenges in ensuring basic health service coverage for their own citizens [3]. Two tier healthcare provision has emerged in Malaysia, with private services limited to those who can afford it and public services for the rest of the population [14]. Thailand's public to private health worker brain drain has strained public health provision, especially in rural areas [15, 16]. Trade in medical supplies, organs, pharmaceuticals and health worker migration have dominated policy debates about the impact on health systems in developing countries, including concerns about intellectual property rights and access to affordable drugs, the latest medical technology, and retaining doctors and nurses within the public sector and/or within the country's health system at all. There are growing concerns about the impact of medical tourism on health systems, particularly equity of access for both foreign and local consumers [17]. Inequities at home, either by low quality services and/or inability to pay, prompt people to seek cheaper and high quality care treatment overseas. As Blouin (2010) contends, a policy question that remains unanswered is whether medical tourism can improve the capacity of poor people in developing countries to access health services. She calls for the exploration of policy mechanisms that mitigate the risks associated with medical tourism, whilst harnessing the potential benefits, for local consumers [18].

In the academic literature, conceptual analyses of medical tourism have emerged from a tourism management perspective, analysing supply and demand factors [19, 20, 21, 22], and as a node in the trade in health perspective [10, 23, 24, 25, 26]. Legal literature is beginning to cover patient liability issues when surgery is carried out overseas [27]. Recent work has begun to analyse medical tourism and its potential impact on health systems in specific countries [1, 28, 29]. Yet not all health systems functions are analysed in these accounts. A core concern is whether medical tourism diverts resources from public components of health systems in destination countries [30]. Furthermore, conceptual frameworks in the health systems literature focus on the impact of targeted, vertical interventions in health systems [31]. But medical tourism is a phenomenon rather than an intervention; its policy implications have yet to be considered within the context of a health system.

This paper presents a conceptual framework of medical tourism and policy implications for health systems in Southeast Asia, drawing on the cases of

Thailand, Singapore and Malaysia, via an extensive review of the academic and grey literature, as well as insights from health consultancies in the public and private sectors across the region. This framework provides a basis for more detailed country specific studies on the benefits and disadvantages of medical tourism, of special relevance for policymakers and industry practitioners in other Southeast Asian countries with similar health systems where governments have expressed interest in facilitating the growth of the medical tourist industry. Bridging the social science disciplines, the public policy approach to research is a pragmatic one, with the end goal of translating research into useful policy recommendations, in this instance those that optimise the benefits of medical tourism for both foreign and local consumers and mitigate the risks. Research methodology is outlined below, followed by the policy implications of medical tourism for health systems at their governance, delivery, financing, human resources and regulation functions [32, 33]. The conclusion emphasizes the need for concerted data collection efforts and identifies variables for further analysis of medical tourism's potential impact on health systems.

2. RESEARCH METHODOLOGY

Media reports on the medical tourism industry and participation in regional conferences enabled the researchers to pinpoint Singapore, Thailand and Malaysia as the three main hubs for medical tourism in Southeast Asia for comparative analysis. Broadly, there are four types of comparative health policy analyses. The first constitute descriptive studies, with no hypothesis or testing of explanations on why patterns exist, leaving policy explanations implicit for the reader to gauge. The second include collections of international case studies with some assessment of performance, whilst the third type includes studies employing a common framework for analysis (e.g. privatization). The fourth type of cross national studies are those that show a fundamental theoretical orientation, with a specific theme or question as a focus of analysis (Marmor et al 2005: 341 - 2) [34]. We decided to undertake this fourth type of comparative analysis, in order to generate a conceptual framework that could be usefully employed by policymakers to understand the policy implications of medical tourism on health systems with similar structures. Methods employed focussed on conceptualising rather than describing, where one or more new concepts are developed to explain what is being studied [35]. An inductive, theory building approach [36] is appropriate to examine medical tourism where knowledge is far lacking, especially in relation to health systems.

An initial informal literature scan using the search criteria "medical tourism AND Asia" in google scholar revealed a lack of data and authoritative sources on medical tourism, particularly figures for number of patients and estimated earnings. Academic literature was searched exhaustively in the PubMed and Social Science Research Network databases using the search criteria "medical tourism AND Asia" (92) and "medical travel AND Asia" (806), generating a range of mostly conceptual research. Abstracts were scanned for reference to Thailand, Singapore and Malaysia and/or reference to health systems in general. Additional articles were located using the reference list of selected articles. Study selection was not systematic; no article was omitted but considered in the context of health systems/medical tourism in Asia (43). Articles gathered were then categorised according to content focus (e.g. privatisation of health systems, medical tourism empirical evidence, health and trade nexus). Following categorisation, all articles were analysed to identify medical tourism interaction points across the health system functions, with new material continually brought into the analysis. Concurrent to the theory building process, quantitative data on the nature of health systems in the three study countries were retrieved from official country sources and the World Health Organization. These data were triangulated with the academic literature to validate claims made about the nature of health systems. This data also enabled the researchers to make systematic comparisons between the three country health systems. Following this step, grey literature were searched using the above search criteria in Factiva, a news item database, to provide examples of recent developments in the medical tourist industry in the three study countries. Other grey literature sources included management consultancy research reports, working papers on medical tourism, and medical tourism industry player's statistics and promotional materials. Subsequent to analysis and identification of the conceptual framework, potential policy options were outlined based on the literature and/or innovative examples of comparative health policy responses in the region. We anticipated that the different nature of health systems (e.g. mostly public versus private delivery) would also generate differential policy implications according to local context. In the course of our comparative analysis, we found this to be the case to a large extent; however, medical tourism poses potential risks and benefits regardless of the current nature of a health system. As a phenomenon, it can fundamentally change the nature of health systems themselves without policy intervention (e.g. shift towards a dominantly private hospital sector). Thus, the policy implications described are broadly applicable to health systems in general, but of particular relevance to policymakers and industry practitioners in other Southeast Asian countries where governments have expressed an interest in developing the medical tourist industry.

3. RESULTS

3.1 Governance in separate domains of trade and health

Medical tourism straddles the policy domains of trade and health. Its rise is situated within the rapid growth of trade in health services, driven by increased international mobility of service providers and patients, advances in information technologies and communications, and an expanding private health sector [10]. Trade by definition is international, but health systems (financing, delivery and regulation) remain nationally bounded. Additionally, trade objectives of increased liberalisation, less government intervention and economic growth generally do not emphasize equity, whereas health sector objectives like universal coverage do. Consequently, actors in the trade and health policy spheres tend to have conflicting objectives, and trade and health governance processes remain relatively separate at three levels; the international (World Trade Organisation (WTO) and World Health Organisation (WHO)), regional (Association of South East Asian Nations (ASEAN)) and national (government ministries). Reconciling the aims of economic growth with equitable health service provision and access makes governance of medical tourism within a country's health system challenging at best and contradictory at worst.

At the international level, there are clear tensions between the goals of protecting and promoting health and generating wealth through trade [23]. Trade and health policy negotiations occur in isolation, despite the growing importance of the trade and health nexus at the global level, e.g. extensive health worker migration and cross border consumption of health services (medical tourism) [10, 23]. WTO membership requires adherence to a multitude of legally binding obligations, including removal of tariff and non tariffs barriers on goods and services. The WTO's formal governance architecture is embodied in its legally binding trade agreements and compulsory legal dispute mechanism. These legal apparatus afford it more compliance clout than the WHO, which by contrast is an advocacy organization. The WHO imposes no legal obligations on members, relies on non binding agreements, and has no compulsory dispute mechanism. Thus enforcement capacity in cases of non compliance to WHO agreements is limited [23]. Economic growth and trade considerations are likely to surpass health objectives at the global level when countries face sanctions or legally punitive measures for non compliance with trade agreements. Examples of trade and health policy incoherence include patents on essential medicines and tobacco promotion in developing countries, permitted by trade agreements [37].

Whilst most trade in health services takes place outside the framework of existing trade agreements, whether bilateral or multilateral [25], trade in health services including medical tourism is officially provisioned for under the General Agreement on Trade in Services (GATS). The four modes of supply include; 1. The cross border supply of services (remote service provision, e.g. telemedicine, diagnostics, medical transcriptions), 2. Consumption of services abroad (medical tourism, medical and nursing education for overseas students) 3. Foreign direct investment (e.g. foreign ownership of health facilities) and 4. Movement of health professionals [7]. Countries can choose to make GATs commitments (which legally bind them to open markets under the auspices and protection of the WTO) sectorally or via a specific mode. In ASEAN, only Cambodia, Malaysia and Vietnam have made GATs commitments relevant to the health sector [38]. Medical tourism is becoming bureaucratized, formalized and normalized [17] evidenced by GATs provisions for the health sector. In the context of increasing cross border trade in health services, governments have the option to either schedule GATs commitments in health or continue to trade outside of formal agreements. With rapidly changing domestic and international health markets, the latter looks likely, but it is worth noting that GATS commitments can also limit the degree to which foreign providers can operate in the market [39]. In policy terms, this clause can protect health systems from monopolization by foreign investors in the health sector.

Regionally, trade also tends to trump health in terms of policy action. ASEAN is primarily a trade forum, and the 1995 ASEAN Framework on Agreement on Trade in Services (AFAS) makes provisions for services liberalisation between members beyond the WTO GATs. Unlike the WTO, ASEAN has no legal authority to enforce compliance, but a dispute settlement mechanism was recently signed. Whilst the health sector is not covered under the AFAS, it is envisioned that the free flow of all goods, services, investments, capital and skilled labour will be achieved to create an ASEAN Economic Community (AEC) by 2020 [40, 41]. The ASEAN Economic Community (AEC) council meets bi annually to work towards deepening and broadening regional economic integration. In contrast, the ASEAN Health Minister's Meeting (AHMM) is held every two years. Currently, ASEAN health cooperation is limited to disaster preparedness for natural disasters and infectious disease outbreaks. Agreements in health are limited to sanitary and phytosanitary measures, bar a non legally binding Mutual Recognition Agreement (MRA) on the movement of health professionals. The ASEAN Work Plan on Health Development (2010 - 2015) was finalised in July 2010 to cover broader regional health issues, including non communicable diseases, maternal and child health and primary health care [42, 43]. Despite ASEAN's regional economic and health integration, there have been no agreements signed concerning the medical tourism industry. Foreign direct investment by regional players in neighbouring countries is accelerating, with private companies like Singapore's Parkway Holdings (one of the largest hospital operators in Asia) and the Raffles medical group acquiring hospitals in Singapore, Malaysia, Brunei, India and China [26]. Malaysia's state investment company Khazanah's $2.6 billion bid in Parkway Holdings in 2010 gave it a 95% stake in the company [44]. Foreign investment by both private and state investment companies implies that significant profits can be made in the health sector of other countries, with profits accruing to shareholders overseas and few benefits for local consumers, unless profits are taxed and reinvested in the destination health system. The substantive economic capacity of these regional players means that health policy aims, like universal access to healthcare, are likely to come secondary to trade policy aims, like increasing foreign investment that can be gained from medical tourism.

Trade and health policy incoherence in promoting both medical tourism and universal coverage for local consumers at the national level is evident. Whilst several studies on medical tourism allude to government's role in promoting medical tourism [8, 16, 21], these do not differentiate between the role of *different* government ministries and their respective policy aims. Trade and tourism ministries are primarily concerned with increasing economic growth and facilitating international trade in the services sector. In contrast, a health ministry's aim is to improve overall population health and ensure equity in health service access and delivery. Health systems are also nationally bounded; maximising scarce public resources for health within given territorial constraints gives rise to healthcare protectionism by governments, typified by strict eligibility requirements for access to state subsidised services by migrants. Whilst expansionist medical tourism policies had been initiated in trade and tourism ministries of all three countries, there appears to be a spill over effect on ministries of health (MOH). Increasingly, MOH's are establishing medical tourism committees and departments, dedicated to the promotion of their respective countries' health facilities to other governments/foreign patients. For example, Thailand's medical hub policy was initiated in 2003 by the government agency the Thailand Board of Investment, whilst the Ministries of Commerce, Department of Export Promotion and the MOH in collaboration with private hospitals are now the main implementers of the policy [15]. Whilst Malaysia's national health plan does not mention medical tourism as a strategic aim [45], the MOH formed an inter-ministerial committee for the promotion of medical and health tourism (MNCPHT) in 2003 [28]. Of the three countries, Singapore's government agencies have the most integrated policy stances that strongly support medical tourism [2], reflective of the country's prioritisation of economic growth. Singapore's Tourism Board, the Ministry of Trade and Industry's Economic Development Board and the MOH have set a target to attract 1 million foreign patients by 2012 [46], whilst one of the MOH's explicit priorities is to "exploit the (country's) economic value as a regional medical hub" [47]. In 2004, a multiagency government initiative (including the MOH) SingaporeMedicine was launched with the aim of developing Singapore as a medical hub. Whilst trade and tourism and health ministry objectives are not easily reconciled, medical tourism growth provides an opportunity for inter ministry policy coordination, e.g. via a cross subsidization mechanism whereby medical tourist revenues are taxed, providing extra income for public hospitals. In the three countries, an apparent convergence in trade, tourism and health ministry priorities is taking place, reflective of growing acceptance of health as a private good globally. Improved data collection on medical tourist flows and health systems use and access by local consumers are necessary to assess whether policies that promote medical tourism and universal coverage are reconcilable. Pre-emptively, government ministries should work towards more integrated governance of medical tourism, especially given the highly privatised health system landscape and existing inequities in health systems use and access by local consumers, which could be aggravated by foreign patient inflows.

3.2 Delivery in private versus public sector

Medical tourism is driven by the for profit private sector in health systems. The private sector dominates primary care provision in Singapore and Malaysia, but is slowly expanding its role in tertiary hospital care. Private primary care providers are concentrated in urban areas, with public primary care providers catering to those in rural areas, as seen in Thailand and

Malaysia [14, 48]. Hospital services are dominated by the public sector, with a 70 - 80% share of beds (table 3) but private hospital providers are steadily growing. In Thailand, private hospital numbers have hovered consistently at 30% of total hospitals between 1994 and 2006 [48]. In Singapore, private sector hospital growth has risen in proportion with public sector hospital growth between 1998 and 2008 [49]. Private hospitals are smaller in size and tend to be located in urban areas, serving middle to high income patients as well as foreign patients [50]. In general, the public private mix of healthcare provision in this region reflects the country's level of economic development. During economic growth periods, wealthier populations have emerged with demand for private providers in response to perceived lower quality public provision. Consequently the public sector has become more pro poor as this group cannot afford private care, leading to the development of a two tier healthcare system seen in Thailand and Malaysia [14, 51]. Public services are generally perceived to be of low quality or unresponsive in this region by local consumers [5052]. The steady growth of private sector hospitals has mirrored the increase in medical tourism (tables 2 and 3).

The link between a growing private, for profit sector that caters to medical tourists and access to such services by local consumers without the ability to pay is elusive. Private ownership of health facilities means that benefits accrued (profits from service fees for foreign patients) are remitted offshore to companies based in different countries who are investing in private hospital chains across Southeast Asia. For example, the recent Fortis-Parkway merger of the second largest Indian healthcare group with the largest private Singapore-Malaysia group created the largest hospital chain in Asia. Parkway's subsequent take-over bid by Malaysia's state investment company Khazanah, means that profits accrued are remitted to Malaysia for health services rendered in Singapore and India. Purchase of costly technology that doesn't have a wider social benefit for the procedures that medical tourists demand has raised concerns about "crowding out" local consumption of high technology procedures [12]. Furthermore, government subsidies for private sector growth, via tax breaks and preferential access to land, is unlikely to benefit the health system at large nor facilitate broader public health goals (universal coverage) if private hospitals cater to larger shares of fee paying, foreign patients. This can be seen in Malaysia, where tax incentives are available for building hospitals (industry building allowance), using medical equipment, staff training and service promotion (deductions on expenses incurred) [8]. Private sector growth in health is implicitly encouraged via these benefits, at the same time as government construction of new hospitals has stalled due to alleged insufficient public funds [56].

Table 3. Public versus private health provision [49, 53, 54, 55]

	Hospitals		Beds		Beds per 1000 population	Primary care clinics	
	Public (%)	Private (%)	Public (%)	Private (%)		Public	Private
Thailand	67.9% (2007)	32.1% (2006)	69.3% (2006)	30.7% (2006)	2.2 (2002)	80.5% (2007)	19.5% (2006)
Singapore	63.6% (2009)	36.4% (2009)	80.6% (2009)	19.4% (2009)	3.2 (2007)	1.5% (2005)	98.5% (2005)
Malaysia	40.6% (2008)	59.4% (2008)	77.9% (2008)	22.1% (2008)	1.8 (2007)	32.1% (2008)	67.9% (2008)

Medical tourism is emerging in public sector hospitals at the same time as it is being driven by the private sector, notably in corporatized (public) hospitals. Corporatization of hospitals in Singapore since 1985 granted hospitals greater autonomy and exposure to market competition under government ownership, with the aim of lowering costs and improving service quality [57]. All public hospitals in Singapore are Joint Commission International (JCI) accredited [58]. Given that these hospitals are publicly owned, revenues accruing to medical tourism are taxable and thus profits can be reinvested back into the public health system by the government. In Malaysia and Thailand, some public hospitals are allowing their surgeons to operate a private wing for private patients, including medical tourists. This policy move could incentivise surgeons to treat the additional fee paying foreign patients over local consumers, when public health resources are already strained in those countries.

The majority of medical tourists in Southeast Asia hail from neighbouring countries, reflecting inequities in service provision at home, either via unavailability of quality services or underinsurance. In Singapore and Malaysia, most medical tourists are from ASEAN countries, whilst Thailand's consumers are often from outside the region, with the Japanese accounting for the largest share of foreign patients (table 2) [50]. Indonesians travel to Singapore and Malaysia for medical treatment, whilst Cambodians cross the border to Vietnam for higher quality health services. Low quality public and private health provision at home forces them to leave for overseas treatment. Cost is a factor, but Malaysian, Singaporean and Thai hospitals offer specialised services unavailable in other, especially poorer, ASEAN countries [2, 50]. The policy implications go beyond the potential to crowd out consumption by locals. As Chee (2010) points out, when middle class fee paying patients decide to undertake treatment abroad, their domestic health systems lose out, not only financially but in terms of the political pressure that these potential consumers could exert to improve the health system that poorer consumers rely upon [28]. The possibility to "exit" low quality health systems gives the middle class little incentive to exert pressure for quality improvement [59]. Policy options that raise quality standards and minimize quality differentials, both within and between countries in Southeast Asia, would benefit both foreign and local consumers. These include public private linkages via professional exchanges, joint training initiatives, shared use of facilities between public and private providers to maximise resource use, telemedicine, and use of complementary/specialised treatments [1, 12].

3.3 Healthcare financing and consumerism

Consumer driven healthcare is becoming the normalised globally and in this region, partly encouraged by governments and the private sector seeking to shift responsibility for one's health to the individual in response to rising healthcare costs and demand for services. Singapore and Malaysia exemplify this trend, as public health expenditure has slowly been declining whilst private health expenditure has increased [28]. The Thai government spent almost double the amount on health as a percentage of total government expenditure (14.1%) compared to Singapore (8.2%) and Malaysia (6.9%) in 2008 [53]. As table 4 shows, the Thai government contributes the majority of total health spending (75.1%), in contrast to Malaysia and Singapore, where private health spending surpasses government health spending. Although both Singapore and Malaysia in theory offer 100% population coverage, high out of pocket payments (OPPs) suggest effective coverage is less than this [52]. Both countries are encouraging greater use of individual financing instruments to pay providers, in addition to compulsory state insurance schemes (Medishield in Singapore) or taxation (Malaysia). These include medical savings accounts (Medisave in Singapore, Employee Provident Fund Account 2 in Malaysia) [60] and widespread private insurance. Thailand is the exception, where the government's commitment to enrolling the population in its universal social insurance scheme means that government investment in health has risen since 2002 [56, 61, 62].

The most regressive financing mechanism, out of pocket payments (OPPs), dominates private health spending in all three countries. More OPPs for services leads to more competition in private healthcare markets, as providers are more likely to compete for patients based on price, especially given the price transparency made possible by the internet. Medical tourist payments are dominated by OPPs, but these payments are becoming more organized as part of insurance coverage. For example, since March 2010 Singapore's Medisave can be used for elective hospitalizations and day surgeries in hospitals of two partner providers in Malaysia, Health Management International and Parkway Holdings [63]. Deloitte's 2009 medical tourism industry report highlighted four US health insurers who are piloting health plans that permit reimbursement of elective procedure overseas in Thailand, India and Mexico [64]. The trend of insurance companies and employers turning to foreign medical providers to reduce costs looks set to continue as the medical tourism industry grows [29].

Table 4. Health expenditure [53]

	Total health expenditure as % of Gross Domestic Product (2008)	Government expenditure on health as % of total government expenditure (2008)	Government health expenditure as % of total health expenditure (2008)	Private expenditure as a % of total health expenditure (2008)	Out of pocket expenditure as a % of private health expenditure (2008)	Private prepaid plans as a % of private health expenditure (2008)
Thailand	4.0%	14.1%	75.1%	24.9%	71.1%	20.9%
Singapore	3.4%	8.2%	35.0%	65.0%	93.9%	2.8%
Malaysia	4.3%	6.9%	44.1%	55.9%	73.2%	14.4%

One policy implication of the increase in medical tourists on health financing is that differential pricing for foreign patients could drive up costs of services for local consumers over time. Redistributive financing mechanisms may offset these increases. Policy options include taxing medical tourist revenues to be reinvested in the public health system [12], expanding financing instruments that do not tie access to ability to pay (taxation, social insurance) and mandating private providers to participate in schemes that provide coverage to local consumers. Private hospitals could provide services to a specified percentage of foreign patients and local consumers enrolled in state schemes, or provide certain specialist treatment for locals (depending on a centre's area of clinical expertise). The need for such policies is pressing when, for example, private hospitals treating foreign patients in Thailand currently do not participate in social health insurance schemes, which covered 98% of the population in 2009 [25, 52, 65].

3.4 Human resources and specialists

Health worker shortages persist to varying degrees in Southeast Asia, at the same time as demand for health services from foreign patients is rising. Whilst all three countries have health worker densities above the WHO critical threshold of 2.28 health workers per 1000 population, all countries face pressures to supply trained health workers to meet population health needs [66, 67]. There are low doctor-to-patient ratios in Thailand and Malaysia (table 5), as well as continual outmigration of doctors from Singapore and Malaysia. Within ASEAN, these two countries record the highest levels of doctor outmigration to OECD countries [68]. International outmigration from Thailand is low, but intra-country migration from rural to urban areas and maldistribution of health workers is common [15, 16]. In response to shortages, Singapore has been able to attract health workers from the Philippines and Malaysia. In Thailand, health workers must pass medical exams in Thai, limiting potential for physician immigration to the country. Whilst the foreign medical workforce inflow to Malaysia has been substantial, this has been insufficient to offset the outflow of Malaysian doctors to other countries [25].

Table 5. Human resources for health [49, 53, 69]

	Doctors per 1000 population	Doctors		Nurses per 1000 population	Nurses	
		Public (%)	Private (%)		Public (%)	Private (%)
Thailand	0.4 (2000)	78.4% (2005)	21.6% (2005)	2.8 (2000)	87.8% (2005)	12.2% (2005)
Singapore	1.5 (2003)	54.8% (2009)	45.2% (2009)	4.5 (2003)	68.5% (2009)	31.5% (2009)
Malaysia	0.7 (2002)	60.1% (2008)	39.9% (2008)	1.8 (2002)	71.2% (2008)	28.8% (2008)

Rising demand for health services in the region has precipitated the growth in private medical and nursing schools across Southeast Asia and correspondent rise in trained health workers. Public and private medical schools in the region are establishing partnerships with reputable universities overseas. Thailand's Mahidol university nursing department has established links with nursing schools in Sweden, Canada, Australia, Korea, the UK and the USA to facilitate student and teaching exchanges. Singapore's National University recently opened a graduate medical school with Duke university in the USA, and Malaysia's Sunway university medical school trains students in partnership with Monash university in Australia. Such partnerships facilitate capacity building in human resources for health, as well as access to new markets for universities overseas. Importantly, these partnerships signal *quality* of human resources, crucial to the promotion of medical tourism [17].

Developing the medical tourism industry can be seen as a tactic to reduce international emigration of health workers, particularly of specialists. Anecdotal evidence from Thailand indicates that medical graduates, having acquired specialised medical degrees abroad, are finding it lucrative and more satisfying to stay in their home country [2]. Politicians in Singapore have reasoned that in order to recruit and retain specialists in a country with a small local population, that the country must attract a high volume of medical tourists. However, within countries, the growth of medical tourism may exacerbate public to private sector brain drain, notably of specialists who provide elective surgeries demanded by foreign patients. Whilst the proportion of doctors working in the public sector is higher than in the private sector in medical tourist countries (table 5), dual practice, whereby doctors combine salaried, public sector clinical work with fee for service private clientele [70], is common amongst specialists in Thailand and Malaysia. Retaining public sector specialists has become a challenge with the prospect of higher salaries and lower workloads in the private sector. Singapore has managed to maintain competitive public sector salaries, but in Thailand and Malaysia, with larger public - private pay discrepancies, medical tourism has the potential to further incentivise specialists to shift to the private sector. Evidence from Thailand suggests that medical tourism is not negatively impacting the health system by pulling doctors from rural areas. Rather, specialists from teaching hospitals in urban areas are shifting to private hospitals catering to foreign patients [67,71]. All three countries have a high number of doctors with specialty training e.g. 77.5% in Thailand in 2006, [48]. But these specialists are concentrated in the private sector; in Malaysia, only 25 - 30% of specialists work in the public sector [72]. Singapore is the exception, where 65% of specialists are in the public sector [73]. The type of surgery matters; for local consumers seeking specialist, essential surgery (e.g. cardiac, transplantation procedures), paying to see a specialist in a private hospital may be the only option. High quality, specialised care is typically provided in private hospitals and can only be afforded by middle to high income patients [50].

Medical tourism could exacerbate already endemic public to private brain drain in the region. A related concern in Thailand is that medical education is largely publicly funded; private hospitals do not share the costs of such education, yet hire from the same pool of graduates as the public sector [50]. Policy options to mitigate internal brain drain include instituting capitation payments for health costs and standard fees for doctors, regardless of whether a patient is local or foreign. Offering higher salaries in the public sector and bonding publicly funded graduates are options for governments (all three countries bond their graduates for between 3 to 5 years). Dual practice of specialists could be allowed but regulated, so that specialists dedicate a specified amount of time to treat local consumers. When public funds are used to train specialists who then shift to the private sector (potentially to treat medical tourists), redistributive government regulations like paying a fee to leave the public sector (Thailand) may plug a short term financial resource gap, but recruitment and retention is a persistent problem in this region.

3.5 Regulation of quality control and new actors

Private hospitals in the three countries are accredited via different channels, leading to differing quality standards between public and private hospitals. Private hospital associations encourage industry self regulation, whereas public hospitals are regulated by the MOH or quasi governmental bodies. For example, publicly owned corporatized hospitals in Singapore operate with autonomy in a competitive environment, but government ownership allows them to shape hospital behaviour without cumbersome regulation [74].

Joint Commission International (JCI) is the most established medical tourist industry accreditor worldwide. Of the three profiled countries, Singapore has the highest number of JCI accredited providers (18), followed by Thailand (13) and Malaysia (7) [58]. JCI accreditation is an important quality signal to attract medical tourists, but this process is voluntary. The differing quality accreditation channels at the national (private hospital associations vs. MOH) and international levels may lead to inequitable quality standards between the public and private sectors, whereby private hospital standards surpass those in public hospitals, reflective of the current situation in low to middle income countries in Southeast Asia. This has implications for the quality of care received by local consumers without the ability to pay for private services, and the potential divergence of health outcomes between private fee paying patients (foreign and local) and those that can't afford such services. Malaysia's Society for Quality in Health (MSQH), a joint regulatory body launched by the Ministry of Health, Association of Private Hospitals of Malaysia and the Malaysian Medical Association, was recently awarded international accreditation by the ISQua on par with JCI. As the MSQH covers both public and private hospitals, this kind of international standard setting for both sectors could provide a regulatory template for other countries pursuing medical tourism, in order to ensure that both local and foreign consumers enjoy similar quality standards. Policy options include common standards for public and private providers [1] regulated by government, as well as compulsory JCI accreditation for hospitals catering to medical tourists.

New brokers that arise between hospitals and patients are proliferating rapidly. These agencies are located in developed and developing countries, connecting prospective patients to providers via the internet. As yet, the medical brokerage industry has no codes of conduct, and the lack of medical training of brokers raises questions about how these new actors evaluate quality of care when choosing which facilities to promote to prospective patients. There are also no explicit formal standards when establishing referral networks, which could be open to abuse, e.g. financial incentives for brokers from providers to promote facilities) [17]. Regulating medical tourist brokers should be a policy priority in both source and destination countries.

4. DISCUSSION AND DIRECTIONS FOR FUTURE RESEARCH

Based on the health systems functions of governance, delivery, financing, human resources and regulation [32, 33], the conceptual framework (Figure 1) aims to provide a basis for further empirical studies weighing the benefits and disadvantages of medical tourism for health systems, of particular relevance to countries in Southeast Asia.

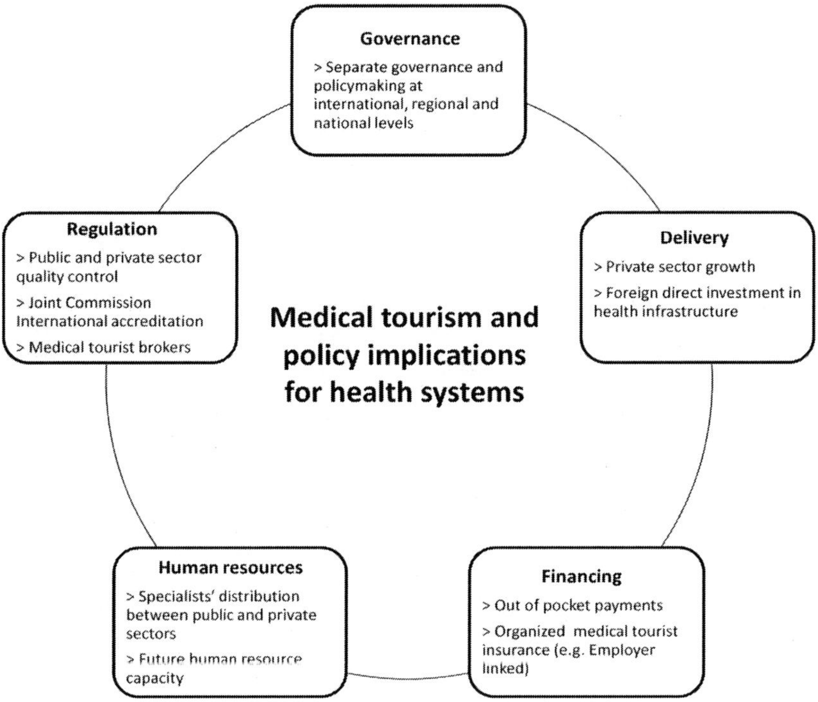

Figure 1. Conceptual framework for medical tourism and policy implications for health systems.

The framework facilitated the identification of the following variables for empirical analysis:

Governance: the number and content of GATs health sector commitments, the number and size of medical tourist government committees or agencies, availability of medical tourist visa.

Delivery: number of hospitals in public and private sector treating foreign patients, consumption of health services by domestic and foreign population (hospital admissions).

Financing: medical tourist revenues, type of medical tourist payment (service fee or insurance, level of copayment), foreign direct investment in the health sector.

Human resources: doctor and nurse ratios per 1000 population, proportion of specialists in the public and private sectors, number of specialists treating foreign patients.

Regulation: number of JCI accredited hospitals, number of medical tourist visits facilitated by brokers.

At present there is an acute lack of reliable empirical data concerning medical tourist flows. Most urgently, a universal definition of who counts as a medical tourist (e.g. per procedure or per inpatient) should be agreed on, ideally at the international (WHO) or regional level (amongst Ministries of Health, Trade, Tourism and private hospital associations). Variation in definitions and estimates amongst the three study countries alone are significant. Singapore's Tourism Board estimates medical tourist inflows based on tourist exit interviews with a small sample population, whilst the Association of Private Hospitals in Malaysia collects data only from member hospitals and includes all foreign patients, including foreign residents and those who happen to require medical care whilst on vacation [28]. Thailand's Ministry of Commerce collects data on medical tourist inflows from private hospitals, counting foreign patients as the APHM does, except that definitions between hospitals about numbers vary (some count inpatient admissions, others per procedure) [71]. Standardised data collection will enable researchers to make meaningful cross country comparisons, as well as carry out detailed country specific studies to investigate the benefits and disadvantages of medical tourism's impact on health systems.

5. CONCLUSION

The rise of medical tourism in Thailand, Singapore and Malaysia and governments' endorsement of the trend has raised concerns about its potential impact on health systems, namely the exacerbation of existing inequitable resource distribution between the public and private sectors. Nowhere is this more evident than in Southeast Asia, where regulation and corrective policy measures have not kept pace with rapid private sector growth during the past few decades. This paper presents a conceptual framework (Figure 1) that identifies the policy implications of medical tourism for health systems, from a comparative analysis of Thailand, Singapore and Malaysia. This framework can provide a basis for more detailed country specific studies, of particular use for policymakers and industry practitioners in other Southeast Asian countries where governments have expressed an interest in facilitating the development of the industry. Medical tourism can bring economic benefits to countries, including additional resources for investment in healthcare. However, unless properly managed and regulated on the policy side, the financial benefits of medical tourism for health systems may come at the expense of access to and use of health services by local consumers. Governments and industry players would do well to remember that health is wealth for both foreign and local populations.

ACKNOWLEDGEMENTS

All analysis and opinions expressed in this paper are the authors' alone. The authors acknowledge the insights on methodology provided by Wu Xun, as well as the detailed and helpful comments from Chee Heng Leng on an earlier version of this draft.

AUTHORS' CONTRIBUTIONS

NP conducted the research and wrote the first and final versions of the draft. KHP commented on the first and subsequent drafts. NP revised the final manuscript. Both authors read and approved the final manuscript.

REFERENCES

1. Hazarika I: Medical tourism: its potential impact on the health workforce and health systems in India. Health Policy and Planning. 2010, 25: 248-251. 10.1093/heapol/czp050.

2. UNESCAP: Medical travel in Asia and the Pacific: challenges and opportunities. 2007, Thailand, United Nations Economic and Social Commission for Asia and the Pacific, Accessed 29-09-2010, [http://www.unescap.org/ESID/hds/lastestadd/MedicalTourismReport09.pdf]

3. Chongsuvivatwong V, Phua KH, Yap MT, Pocock NS, Hashim JH, Chhem R, Wilopo SA, Lopez AD: Health and health-care systems in southeast Asia: diversity and transitions. Lancet. 2011, 377 (9763): 429-37.

4. Tourism Malaysia: Media info: Health tourism in Malaysia. 2008, Communications and Publicity Division, Tourism Malaysia, Accessed 29-9-2010, [http://www.tourism.gov.my/corporate/images/media/features/Health%20Tourism%20Oct%202008.pdf]

5. Singapore Medicine: Media Factsheet: Singapore, more than just a world class healthcare destination. 2010, Singapore Medicine

6. Carrera P, Bridges J: Health and medical tourism: what they mean and imply for health care systems. Health and Ageing. 2006, The Geneva Association, 15:

7. Carrera P, Bridges J: Globalization and healthcare: understanding health and medical tourism. Expert Review of Pharmacoeconomics and Outcomes Research. 2006, 6 (Suppl 4): 447-454.

8. Chee HL: Medical tourism in Malaysia: international movement of healthcare consumers and the commodification of healthcare. Asia Research Institute, National University of Singapore, submitted. 2007, Accessed 04-03-2011

9. Whittaker A, Manderson L, Cartwright E: Patients without borders: understanding medical travel. Medical Anthropology. 2010, 29 (4): 336-343.

10. Smith R, Chanda R, Tangcharoensathien V: Trade in health related services. Lancet. 2009, 373: 593-601.

11. Blouin C, Chopra M, Van Der Hoeven R: Trade and social determinants of health. Lancet. 2009, 373: 502-507.

12. Chanda R: Trade in health services. Bulletin of the World Health Organisation. 2002, 80: 158-163.

13. Hopkins K, Labonte R, Runnels V, Packer C: Medical tourism today: what is the state of existing nowledge?. Journal of Public Health Policy. 2010, 31 (2): 185-198.

14. Chee HL: Ownership, control and contention: challenges for the future of healthcare in Malaysia. Social Science and Medicine. 2008, 66: 2145-2156.

15. Pachanee C, Wibulpolprasert S: Incoherent policies on universal coverage of health insurance and promotion of international trade in health services in Thailand. Health Policy and Planning. 2006, 21 (4): 310-318.

16. Wibulpolprasert S, Pengpaibon P: Integrated strategies to tackle the inequitable distribution of doctors in Thailand: four decades of experience. Human Resources for Health. 2003, 1 (12): Accessed 04-03-2011, [http://www.human-resources-health.com/content/1/1/12]

17. Turner L: First world health care at third world prices: globalization, bioethics and medical tourism. Biosocieties. 2007, 2: 303-325.

18. Blouin C: Trade in health services: can it improve access to health care for poor people?. Global Social Policy. 2010, 10 (3): 293-295.

19. Connell J: Medical tourism: sea, sun sand and ... surgery. Tourism Management. 2006, 27: 1093-1100.

20. Garcia Altes A: The development of health tourism services. Annals of Tourism Research. 2005, 32 (1): 262-66.

21. Henderson JC: Healthcare tourism in Southeast Asia. Tourism Review International. 2004, 7: 111-121.

22. Heung VCS, Kucukusta D, Song H: A conceptual model of medical tourism: implications for future research. Journal of Travel and Tourism Marketing. 2010, 27: 336-251.

23. Fidler DP, Drager N, Lee K: Managing the pursuit of health and wealth: the key challenges. Lancet. 2009, 373: 325-31.

24. Lee K, Sridhar D, Patel M: Bridging the divide: global governance of trade and health. Lancet. 2009, 373: 416-22.

25. Arunanondchai J, Fink C: Trade in health services in the ASEAN region. Health Promotion International. 2007, 21 (Suppl 1): 59-66.

26. Diaz Benavides D: Trade policies and export of health services: a development perspective. Trade in Health Services, Global, Regional

and Country Perspectives. 2002, Washington DC: Pan American Health Organization, Program on Public Policy and Health, Division of Health and Human Development, 53-69.

27. Cohen G: Protecting patients with passports: medical tourism and the patient protective argument. Harvard Law School Public Law and Legal Theory Working Paper. 2010, 10-08:

28. Chee HL: Medical tourism and the state in Malaysia and Singapore. Global Social Policy. 2010, 10 (3): 336-357.

29. Forgione D, Smith P: Medical tourism and its impact on the US health care system. J Health Care Finance. 2006, 34 (1): 27-35.

30. Pennings G: Ethics without boundaries: medical tourism. Principles of health care ethics. Edited by: Ashcroft RE, Dawson A, Draper H, McMillan JR. 2007, Chicester: John Wiley and Sons, 505-510. 2

31. Atun R, De Jongh T, Secci F, Ohiri K, Adeyi O: Integration of targeted health interventions into health systems: a conceptual framework for analysis. Health Policy and Planning. 2010, 25: 104-111.

32. Badara S, Evans T, Dybul M, Atun R, Moatti JP, Nishtar S, Wright A, Celletti F, Hsu J, Kim JY, Brugha R, Russell S, Etienne C: An assessment of interactions between global health initiatives and country health systems. Lancet. 2009, 373: 2137-69.

33. Hsiao W: What is a health system? Why should we care?. Harvard School of Public Health, working paper. 2003

34. Marmor T, Freeman R, Okma K: Comparative perspectives and policy learning in the world of health care. Journal of Comparative Policy Analysis. 2005, 7 (4): 331-348.

35. Punch K: Introduction to social research: quantitative and qualitative approaches. 2005, London: Sage, 2

36. Glaser BG, Strauss AL: The discovery of Grounded Theory: strategies for qualitative research. 1967, New York: Adline Publishing

37. Zeigler D: International trade agreements challenge tobacco and alcohol control policies. Drug and Alcohol Review. 2006, 25: 567-579.

38. WTO GATs services database. Accessed 10-2-2011, [http://tsdb.wto.org/default.aspx]

39. WTO GATs introduction. Accessed 10-2-2011, [http://www.wto.org/english/tratop_e/serv_e/gats_factfiction1_e.htm]

40. ASEAN: ASEAN cooperation in trade in services. 2000, ASEAN

41. Hew D, Soesastro H: Realizing the ASEAN economic community by 2020: ISEAS and ASEAN - ISIS approaches. ASEAN Economic Bulletin, Institute of Southeast Asian Studies. 2003

42. ASEAN: ASEAN annual report 2009 - 2010: Bridging markets, connecting peoples. 2010, ASEAN

43. ASEAN: Joint statement of the 10th ASEAN health ministers meeting: "healthy people, healthy ASEAN". Singapore. 2010

44. Huang R: Khazanah controls 95% of healthcare firm Parkway Holdings. 2010, Channel News Asia

45. MOH Malaysia: Strategic plan 2006 - 2010. 2008, Ministry of Health Malaysia

46. MTI Singapore: Developing Singapore as a compelling hub for healthcare services in Asia. 2002, Health Sector Working Group, Ministry of Trade and Industry Singapore

47. MOH Singapore: Roles and priorities. 2008, Ministry of Health Singapore, Accessed 14-7-2010, [http://www.moh.gov.sg/mohcorp/about.aspx?id=116]

48. MOH Thailand: Thailand health profile report 2005 - 2007. 2007, Ministry of Health Thailand, Accessed 14-7-2010, [http://www.moph.go.th/ops/thp/index.php?option=com_content&task=view&id=6&Itemid=2]

49. SingStat: Yearbook of Statistics 2009. 2009, Singapore Department of Statistics

50. Arunanondchai J, Fink C: Trade in health services in the ASEAN region. World Bank Policy Research Working Paper. 2007, 4147:

51. O'Donnell O, Doorslaer E, Rannan-Eliya RP, Somanathan A, Adhikari SR, Harbianto D, Garg CC, Hanvoravongchai P, Huq MN, Karan A, Leung

GM, Ng CW, Pande BR, Tin K, Tisayaticom K, Trisnantoro L, Zhang Y, Zhao Y: The incidence of public spending on healthcare: comparative evidence from Asia. World Bank Economic Review. 2007, 21 (1): 93-123.

52. Tangcharoensathien V, Patcharanarumol W, Ir P, Aljunid SM, Mukti AG, Akkhavong K, Banzon E, Huong DB, Thabrany H, Mills A: Health-financing reforms in southeast Asia: challenges in achieving universal coverage. Lancet. 2011, 377 (9768): 863-873.

53. WHO: Global Health Observatory. Accessed 20-01-2011, [http://apps.who.int/ghodata/]

54. MOH Malaysia: Health facts 2008. 2009, Health Informatics Centre, Planning and Development Division, Ministry of Health Malaysia

55. National Statistical Office of Thailand: The 2007 private hospital survey. 2007, Ministry of Information and Community Technology

56. Ramesh M, Xun W: Realigning public and private health care in Southeast Asia. The Pacific Review. 2008, 21 (2): 171-187.

57. Phua KH: Attacking hospital performance on two fronts: network corporatization and financing reforms in Singapore. Innovations in health service delivery: the corporatization of public hospitals. Edited by: Preker AS, Harding A. 2003, Washington DC: Health, Nutrition and Population Series World Bank, 451-485.

58. Joint Commission International: Accredited organizations. Accessed 20-01-2011, [http://www.jointcommissioninternational.org/jci-accredited-organizations/]

59. Hirschman AO: Exit, Voice, and Loyalty: Responses to Decline in Firms, Organizations, and States. 1970, Cambridge, MA: Harvard University Press

60. Ministry of Finance Malaysia, Employees Provident Fund. Accessed 20-01-2011, [http://www.kwsp.gov.my/index.php]

61. Phua KH: Health care financing reforms in the Asia-Pacific region. Public Administration and Policy. 2003, 12 (1): 13-36.

62. Phua KH, Chew AH: Towards a comparative analysis of health system reforms in the Asia - Pacific region. Asia Pacific Journal of Public Health. 2002, 14 (1): 9-16.

63. MOH Singapore: Medisave for approved overseas hospitalization. 2010, MOH press release

64. Deloitte Center for Health Solutions: Medical tourism: update and implications. Deloitte. 2009, Accessed 20 - 1 - 2011, [http://www.deloitte.com/us/medicaltourism]

65. Hughes D, Leethongdee S: Universal coverage in the land of smiles: lessons from Thailand's 30 baht health reforms. Health Affairs. 2007, 26: 999-1008.

66. WHO: World health report 2006. 2006, Geneva, Switzerland

67. Kanchanachitra C, Lindelow M, Johnston T, Hanvoravongchai P, Lorenzo FM, Huong NL, Wilopo SA, De la Rosa JF: Human resources for health in southeast Asia: shortages, distributional challenges, and international trade in health services. Lancet. 2011, 377 (9767): 769-781.

68. Bhargava A, Docquier F, Moullan Y: New medical brain drain data set. 2010, Accessed 10-2-2011, [http://perso.uclouvain.be/frederic.docquier/oxlight.htm]

69. WHO: World health statistics 2009. 2009, World Health Organisation, Accessed 29-9-2010, [http://www.who.int/whosis/whostat/2009/en/index.html]

70. Ferrinho P, Van Lerberghe WV, Fronteira I, Hipolito F, Biscaia A: Dual practice in the health sector: a review of the evidence. Human Resources for Health. 2004, 2: 14-10

71. Phyu YM, Chotbenjakul W: Medical tourism in Thailand and its impact on the internal brain drain of doctors. Submitted masters thesis. 2010, National University of Singapore, Lee Kuan Yew School of Public Policy

72. Quek D: The Malaysian healthcare system: a review. Intensive workshop on health systems in transition: 29 - 30 April 2009; Kuala Lumpur. 2009, University of Malaya

73. MOH Singapore: Health facts. 2008, Ministry of Health Singapore, Accessed 14-7-2010, [http://www.moh.gov.sg/mohcorp/statistics.aspx?id=5966]

74. Ramesh M: Autonomy and control in public hospital reforms in Singapore. American Review of Public Administration. 2008, 38 (1): 62-79.

Chapter 5

The potential for bi-lateral agreements in medical tourism: A qualitative study of stakeholder perspectives from the UK and India

Melisa Martínez Álvarez, Rupa Chanda and Richard D Smith

[1]Department of Global Health and Development, London School of Hygiene and Tropical Medicine, 15-17 Tavistock Place, London WC1H 9SH, UK.

[2]Indian Institute of Management Bangalore, Bannerghatta Road, Bangalore, India.

ABSTRACT

Background

Globalisation has prompted countries to evaluate their position on trade in health services. However, this is often done from a multi-lateral, rather than a regional or bi-lateral perspective. In a previous review, we concluded that most of the issues raised could be better addressed from a bi-lateral relationship. We report here the results of a qualitative exercise to assess stakeholders' perceptions on the prospects for such a bi-lateral system, and its ability to address concerns associated with medical tourism.

Methods

30 semi-structured interviews were carried out with stakeholders, 20 in India and 10 in the UK, to assess their views on the potential offered by a bi-lateral relationship on medical tourism between both countries. Issues discussed include data availability, origin of medical tourists, quality and continuity of care, regulation and litigation, barriers to medical tourism, policy changes needed, and prospects for such a bi-lateral relationship.

Results

The majority of stakeholders were concerned about the quality of health services patients would receive abroad, regulation and litigation procedures, lack of continuity of care, and the effect of such trade on the healthcare available to the local population in India. However, when considering trade from a bi-lateral point of view, there was disagreement on how these issues would apply. There was further disagreement on the importance of the Diaspora and the validity of the UK's 'rule' that patients should not fly more than three hours to obtain care. Although the opinion on the prospects for an India-UK bi-lateral relationship was varied, there was no consensus on what policy changes would be needed for such a relationship to take place.

Conclusions

Whilst the literature review previously carried out suggested that a bi-lateral relationship would be best-placed to address the concerns regarding medical tourism, there was scepticism from the analysis provided in this paper based on the over-riding feeling that the political 'cost' involved was likely to be the major impediment. This makes the need for better evidence even more acute, as much of the current policy process could well be based on entrenched ideological positions, rather than secure evidence of impact.

1. BACKGROUND

In an increasingly globalised world, countries are assessing their position on trade in health services [1]. Debates on the subject often centre on the World Trade Organisation's General Agreement on Trade in Services [2]. This is removed from the reality, however, where most trade takes place regionally or bi-laterally [3]. In addition, given that there is no systematic collection of data on the quantity of trade in health services that takes place, or the impact it has on the health system [1], discussions on whether to liberalize this type of trade tend to be more influenced by ideology than evidence [4]. Different types of evidence are needed on this debate. These include basic data on trade flows across countries, evidence of the impact of this type of trade on economies and health systems, and a greater understanding of how such trade is perceived, in terms of those involved in it, as barriers or facilitators, and the relative benefits and risks it presents. It is therefore important when seeking to understand and advise a country's position on trade to understand the various perspectives of those involved [5]. To our knowledge, such an analysis of stakeholder perspectives has not been undertaken in the area of trade and health, which this paper seeks to address.

The context for the study is *medical tourism*, as this is one of the highest profile modes of trade in health services (along with health worker migration) [1]. Medical tourism can be defined as the movement of patients across an international border for the purposes of obtaining healthcare. This is usually motivated by long waiting times, high costs of medical care or unavailability of care in the patients' home country [6, 7]. This definition includes both patients paying out of pocket and the potential scenarios where governments or insurance companies pre-arrange foreign healthcare treatment for patients. Although currently the vast majority of medical tourism occurs from patients arranging and paying for the medical services abroad themselves, arguably the biggest *potential* for this market lies within governments or insurance companies pre-arranging foreign care for patients. It is therefore the potential for the development of this latter form of medical tourism that this paper concentrates upon. Currently most medical tourism takes place by individuals privately arranging healthcare in a foreign country. Often, medical tourism facilitators play a role in this by providing information on countries, hospitals and services, and sometimes even arranging the care for the patients. This, however, occurs under general regional or global agreements, which are by definition multi-lateral.

There are no agreed figures for the number of people who go abroad for healthcare, and the data available in the literature are wide-ranging (from thousands to millions) [8]. There have been several reports from consultancy firms which are often quoted to portray medical tourism as a market with great potential, but it is often unclear how their data were generated [9, 10, 11]. Medical tourism raises issues of quality of care [12], litigation [13, 14], continuity of care [15], equity [16] and the creation of an internal brain drain, where medical professionals leave the public health service to work for the private hospitals that cater for (more profitable) foreign patients, further exacerbating staff shortages [17]. Countries need to take these issues into account when deciding on whether (and to what extent) to liberalize trade in health services. However, a missing piece in the discussion thus far has been a focus upon what type of trade relationship they may wish to engage in.

The literature concerning medical tourism is expanding, and there have been several recent attempts to synthesise what is already known about this practice [4, 8, 18, 19, 20]. Although these reviews take different perspectives on medical tourism, they all agree on the potential benefits and dangers it can bring to *importing* and *exporting* countries. Importing countries (those where the medical tourists come from, as these countries are importing healthcare services) can benefit from lower costs and reduced waiting lists; however, there are concerns regarding quality of services and litigation procedures. On the other hand, *exporting* countries (those that provide the health services to foreign patients) can bring in foreign exchange and may prevent health professionals leaving their country to work in overseas institutions, but risk creating a two-tiered system, with foreign patients receiving better care than domestic patients. In addition, reviews highlight the dearth of primary data available on medical tourism, both at the national and international level, and call for more empirical research to be carried out to ascertain the numbers of medical tourists and the procedures they undergo, as well as the validity of the claims made for and against this practice. Finally, the literature highlights the need for more and better regulatory frameworks for medical tourism, such as the creation of specific national medical guidelines to guide the provision of services to foreign patients [21], the adoption of equitable buying guidelines to address the impacts of medical tourism on equity [19], and the development of a common international regulatory platform and reporting system, that goes beyond the current regulatory bodies, such as the Joint Commission International [4]. Nonetheless, these alternative frameworks do not address the principle trade-agreement between nation states, and in this respect this paper adds to the literature by explicitly seeking to gain information relating to the potential for a bi-lateral relationship to overcome some of the core concerns raised thus far in the context of multi-lateral trade agreements (e.g. GATS).

To explore the range of issues, risks and opportunities offered to countries when engaging in bi-lateral - rather than multi-lateral - trade in more depth, the authors conducted a systematic literature review, which is reported elsewhere [Smith R, Martínez Álvarez M, Chanda R: Medical Tourism: a review of the literature and analysis of a role for bi-lateral trade, Submitted]. This literature review focussed on the UK and India as a case study of how the issues highlighted by the literature could be addressed from a bi-lateral perspective. The review supported the findings of the wider literature, as reported earlier, regarding the lack of empirical data on medical tourism and the potential benefits and risks that both importing and exporting countries face when engaging in this type of trade.

The review did, however, conclude that most of the issues that arose could be better addressed from a bi-lateral relationship, where a contract can be drawn out between the two countries outlining, for instance, which hospitals would provide the care, what procedures would be followed should something go wrong, what care the patients would receive before and after going abroad, and how this relationship could improve, rather than damage, the healthcare available to the local population. Nonetheless, there are still questions concerning the perception of such a bi-lateral system, and its ability to address concerns associated with medical tourism.

In an effort to contribute further to this literature-based evidence, the authors thus subsequently carried out semi-structured interviews with key stakeholders in both India and in the UK. The aim was to elicit their views on the prospects and impediments for a bi-lateral trade relationship in medical tourism *per se*, and specifically one between the UK and India as the principle case-study context for discussion (where the UK was seen as an importer and India as an exporter of medical services). This paper summarises the results from these interviews. Following this introduction, the methodology used to undertake and analyse the interviews is provided, followed by the results. The discussion of these results is set within the context of the results from the previous literature review, with the paper concluding with lessons for further research and policy in this area.

2. METHODS

A total of 30 semi-structured interviews were carried out, 20 in India and 10 in the UK. These two countries were chosen because: (i) the UK has some experience with cross-border mobility of patients within the European Union; (ii) India has already made some inroads as an exporter of health services, for instance in the provision of telemedicine services, and is one of the biggest players in medical tourism; and (iii) the two countries have historical ties, a common language, commonalities in the educational system, and there is a large Indian Diaspora population living in the UK, which would facilitate this type of trade. The balance in the number of interviews between the two countries reflects the wider range of stakeholders in India, as a destination for medical tourism, given the variety of the different providers of health services, and the different levels of government and private sector that would be involved in such a decision. The initial participants were selected to represent the key stakeholders likely to be involved in medical tourism as identified through the systematic review mentioned earlier; these were healthcare providers, members of the Department of Health and medical tourists. These stakeholders were accessed through a variety of methods. Healthcare providers and Department of Health officials were selected according to their role. Medical tourists were accessed through personal contacts of the authors. During the interviews, a snowball sampling technique was used to identify others that may have interest and be relevant to the study. The aim of using this type of sampling was to maximise the range of opinions and perspectives and to get as complete a picture on the subject as possible. Recruitment was undertaken until saturation was felt to have been reached (that no new substantive issues were being raised). The stakeholders that took part are listed in table 1. In addition, workshops were held in both India and the UK before and after the interviews took place with officials from the Department of Health, researchers, think tanks, medical professionals and hospital managers. The aim of the workshops was twofold: (i) before the interviews took place, to map out the issues and different perspectives on medical tourism, in order to guide the development of the semi-structured interview schedule; and (ii) after the interviews, to corroborate the findings and obtain any further insights.

Table 1. Stakeholders that took part in the study

Stakeholder group	Country	Stakeholder Number
Department of Health	UK	1, 2, 3,
Ministry of Health	India	20,21
Healthcare provider	UK	4
Healthcare provider	India	13, 14, 16, 17, 18, 19, 22, 23, 24, 25, 28,
Medical tourism facilitator[1]	UK	6
Medical tourism facilitator[1]	India	27
Think tank	UK	7
Medical tourists and companions	UK	8, 9, 10
NGO	India	11
Academic	India	26
Insurance company	India	29,30
Industry association	UK	5
Industry association	India	12, 15

[1]In the category of Medical tourism facilitators we included agencies that either pre-arranged services for the medical tourists (including transport, hospital stay, medical procedures, accompanying person) or provided information services on medical tourism, destinations and providers to potential medical tourists.

The interviews took place between November 2009 and January 2010 and were carried out by one of the authors in the UK (MMA) and another author in India (RC). The interviews lasted approximately one hour, and were semi-structured in nature [22]. This required a qualitative instrument to be designed to guide the interviews; as indicated, this was based on material obtained from the literature review and the first workshops [Smith R, Martínez Álvarez M, Chanda R: Medical Tourism: a review of the literature and analysis of a role for bi-lateral trade, Submitted]. An example is presented in Additional file 1: Appendix A. The instrument was used as a topic guide, but not all questions were covered in all of the interviews, nor were they asked in strictly the order that they are presented. Rather, the instrument was adapted to each participant. The broad themes were:

general perspectives, medical tourism-related activities and operations, and policy issues.

The interviews were conducted face-to-face or over the telephone, and the answers recorded by hand, and then typed up. Written consent was obtained from each interviewee before the interview took place.

The interview transcripts were analysed using thematic [23] and directed content analysis [24]. The scripts were reviewed and themes were identified. These were coded based on the results from the literature review and the workshops carried out previously by the authors. Any additional themes that arose from the transcripts were incorporated (see table 2 for a complete list of coding categories). This was carried out by the first author (MMA), and approved by the other authors (RC and RS). No software was used in the analysis of the interviews.

3. RESULTS

A summary of all the coding categories and the frequency of these by the different groups of stakeholders are shown in table 2. The core findings from this are explored in more detail below.

3.1 Data availability

The review of the literature carried out previously by the authors had shown that there was very little data, and of poor quality, available on medical tourism flows. This is of concern, since in order to determine the scale and impact of medical tourism, data is needed by policy-makers on, for instance, the numbers of medical tourists, the revenues these bring, the outcomes of treatment and the impacts medical tourism has on the home and destination health system. Some of the Indian healthcare providers interviewed estimated that foreign patients made up 5-7% of all patients they treated, with foreign patients representing a slightly higher share of revenue source (although this figure includes foreign patients that were not medical tourists). Many respondents quoted figures from reports carried out by the Confederation of Indian Industries, McKenzie and Deloitte suggesting the promising *prospects* for medical tourism, rather than actual current figures, whilst others, such as medical tourism facilitators, were highly critical of these reports:

Table 2. Coding categories and frequency by groups

Coding Category[1]	Frequency by groups										
	Government officials		Healthcare providers		Medical Tourism Facilitators		Medical Tourists (UK nationals)	Industry associations		Insurance company	Others[1]
	UK	India	UK	India	UK	India		UK	India		
Data				8	1	1					1
Countries		1		10		1			2		2
Regulation	1			8		1			1	1	2
Quality		1	1	4		1	1		1	1	3
Litigation	2		1	2		1	2		2		1
three hour "rule"	2	1		3			2		2		2
Role of government	1			6		1			1		4
Diaspora	2	2	1	6			1		1		3
Local population	1	1		3		1			1		3
Continuity of care	1		1	1			2				1
Prospects	2		1	2	1						1
Perception of India			3		1			1		1	1

[1]The coding categories were identified from the interview transcripts, and influenced by the literature search. They are summarised here. **Data:** Any

data that the participant had available on the size of the medical tourism market, either in global/national terms or numbers from their institution. **Countries:** Participants were asked to name countries involved in medical tourism, both as importers and exporters. **Regulation:** Although not specifically asked for, the lack of regulation of the medical tourism market, and the perceived lack of regulation in health services in the exporting country were mentioned by a significant number of the participants as a key concern.**Quality:** Similarly, the quality of the healthcare services obtained abroad was a key concern for the stakeholders interviewed. **Litigation:** Many stakeholders showed concerns about the differences in malpractice laws and litigation procedures between the exporting and importing countries. **Three hour 'rule':** One of the key barriers highlighted specific to the UK market, the three hour flight restriction on how far patients can travel to obtain healthcare was also highlighted by the respondents. **Role of government:** A bi-lateral agreement between the UK and India on medical tourism would need both governments to be involved; we therefore asked participants on what role (if any) they thought government should have in this. **Diaspora:** The Diaspora appeared to be a key target audience for medical tourism, as the perception was that many would prefer to return "home" for care.**Local population:** Another key concern highlighted in the literature and many of the interviews was the impact medical tourism would have on the healthcare available to the local population. **Continuity of care:** Many of the stakeholders highlighted the lack of continuity of care a patient would experience when going abroad as a major issue. Additionally, some also feared they may cause a bigger burden on the health system if they have complications once they return. **Prospects:** This category included respondents' views on the prospects for a bi-lateral trade relationship on medical tourism between the UK and India. **Perception of India:** Participants' perception of India was a key influence on their views on the viability for this type of trade relationship between the two countries.

Think tank, NGO, Public Health Foundation of India

Medical tourism is a much overhyped market. People have a vested interest saying it is the next big thing, but the numbers are exaggerated ... However, because of these high figures, people see it as a growth market, which results in too many organisations in the market (hospitals, countries, agents) but not enough patients to go round. (Respondent #6, UK medical tourism facilitator)

In general, it was clear that the stakeholders interviewed had very little information on the current status of medical tourism in general, and that related to India and the UK specifically, generating concern for the development of policy concerning this area.

3.2 Origin of medical tourists

Most respondents commented on the target population for the private hospitals in India that attract medical tourists. When asked about which countries international patients came from, most participants from India named neighbouring and low-income countries, with the UK rating low on their priority list. A typical response was that:

The bulk of foreign patients come from Afghanistan, Bangladesh, Nigeria, Iraq, and Ethiopia. The SAARC [South Asian Association for Regional Coordination] *countries account for 50% of all foreign patients, another 20-25% are from Afghanistan, 3.5% or so from the Middle East, 3% from the US ... The UK is not a large market* (Respondent #18, Indian healthcare provider)

There was also widespread agreement, both from the respondents in the UK and in India, that the Diaspora community would be willing to travel back "home" for treatment. However, opinion was divided on the significance of this population. Most of the UK and some of the Indian stakeholders believed that the number of these potential travellers would be small in absolute terms:

There are small numbers of citizens of an Indian origin that would benefit from this type of arrangement, for cultural or religious reasons, but the numbers are small (Respondent #1, UK Department of Health)

Whereas, the majority of the Indian participants believed them to be a key target group for medical tourism:

The main audience could be the NRI [Non-Resident Indian] *population in these developed countries*(Respondent #12, Indian industry association);

Nonetheless, there was widespread agreement between stakeholders that hospitals in India should concentrate on the large domestic market, before targeting foreign patients; this was the view of government officials in both countries, members of academic and industry associations, although, perhaps predictably, only a small proportion of the Indian healthcare providers.

Indian private hospitals argue they have a lot of extra capacity. However, it is hard to believe they are covering the whole of the Indian population (Respondent #5, UK industry association)

The negative part of this is that to chase that foreign segment of patients, we may end up neglecting others needing healthcare in our own country (Respondent #11, Indian NGO)

The medical tourists interviewed did not comment on this issue.

3.3 Quality, perception, regulation and litigation

The majority of the UK stakeholders interviewed showed concerns regarding the quality of healthcare in India, and a potential lack of regulation.

In the UK we have the Care Quality Commission, guidelines, etc., but how do you ensure that the standards are the same or better? (Respondent #4, UK healthcare provider)

However, this was contradicted by most of their Indian counterparts, who believed healthcare providers in India have made great progress in obtaining national and international accreditation, and that the standard of care was very high. They were, though, very aware of the poor perception that the UK had of Indian healthcare.

The main barrier is perception ... but the procedures and staff are of top quality ... Quality is not a problem for Indian corporate hospitals as 70-80% of their doctors are often trained in the UK and the standards are equivalent to those in the US or the UK (Respondent #17, Indian healthcare provider)

Now NABH [National Accreditation Board for Hospitals & Healthcare providers] *is ISQua*[International Society for Quality in healthcare] *accredited. By this certification, the basic level of hospital quality is now ensured in India.* (Respondent #14, Indian healthcare provider)

Nevertheless, this was not the opinion of all Indian stakeholders: whilst most healthcare providers agreed that standards were high and regulation was necessary, some members of academic and industry associations believed the accreditation industry was not as reliable as some believe it to be, and many establishments were wasting their money getting different types of accreditation.

There is concern that now with the NABH and the insistence on it being mandatory, quality may come down over time ... There is also corruption among the auditors and the process of accreditation can be questioned (Respondent #22, Indian healthcare provider)

When discussing quality concerns, most participants linked these to the issue of litigation, which was a concern across the board for Indian and UK stakeholders.

Under a bilateral agreement, care in India would be commissioned by the NHS, and therefore the NHS is automatically responsible for any litigation payments, which would outweigh the savings made (Respondent #1, UK Department of Health)

If one patient sues us, we can have huge problems ... We do not follow systems but have money and influence ... Our legal framework is poor. In the consumer court, the number of pending cases is huge and corruption is also at a high level (Respondent #12, India industry association)

There are concerns about medical malpractice. The company has decided to stay away from issues of liability. There are indemnity clauses built in (Respondent #27, Indian medical tourism facilitator)

Interestingly, the medical tourists that took part in the interviews did not think they had the right to any sort of compensation, only one had ensured that the hospital would take full responsibility and carry out any further procedures for free should anything have gone wrong. One medical tourist had experienced bad quality and potential medical malpractice, but no action was taken.

Many weeks after the surgery she [respondent's mother] *was no better but they* [the doctors in India] *did not listen to us and told us that it would get better in time. My mum's problem has got worse with more complications and the doctors here said the surgery there was not required. We believe that the doctor there used us just to make a profit* (Respondent #10, medical tourist companion).

3.4 Three hour 'rule'

One of the most controversial issues was the three-hour 'rule'. This 'rule' was supposedly devised by the UK Department of Health when considering sending patients for care to other EU countries, as part of EU-wide legislation, stating that patients were not allowed to be treated in countries that were more than three-hours flying time away. The participants from India thought that this was a major formal barrier for the NHS when considering sending patients to India.

The main problem is the three hour restriction by the NHS. This has to be taken up in a bilateral dialogue with the UK on a regular basis and the time has to be increased to eight hours to cover India (Respondent #21, Indian Ministry of Health)

However, the opinion in the UK was divided. Although respondents from industry considered it an important barrier, it was interesting that participants from both the NHS and the Department of Health considered it only "informal guidance":

This is neither a legal rule nor formal guidance. How far someone can travel depends on the clinical needs of the individual. The three hour rule is not a formal binding, and it is not used ... it is up to consultants and PCTs [Primary Care Trusts] *to decide how long someone can travel ... It wouldn't make sense, as there are areas within the EU that are further than 3 hours!* (Respondent #1, UK Department of Health)

The patients interviewed strongly disagreed with the 'rule':

It's a ridiculous rule. If they are willing to consider sending patients to the EU on a case-by-case basis, then they should use that same basis to let them go further. This would not be the case for patients that would need heart surgery, but for things like knee, joints, fingers, eyes, etc. (Respondent #8, medical tourist)

Clearly the status of such a 'rule', illustrative of other perceived barriers in other contexts, is of major significance to establish in any discussions about more formal trade agreements in this area.

3.5 Continuity of care

The problem of continuity of care was highlighted by all groups of participants, both in the UK and in India. In fact, for the UK respondents, it appeared to be a more important issue than the 'three-hour rule'.

If they [medical tourists] *need continuity of care, especially if problems occur, who is going to be there to fix them? There would be an unwillingness here* [UK] (Respondent #4, UK healthcare provider)

Patients also showed concerns about continuity of care, although they did not express any problems in getting care once they returned.

They [domestic healthcare providers] *should at least arrange the aftercare and pre-treatment for those that choose to go privately, as they would be saving money by not having to operate on them*(Respondent #9, medical tourist)

3.6 Prospects and policy changes needed

In general, respondents believed the prospects for a bi-lateral relationship in this type of trade in health services between India and the UK to be low given the present circumstances.

The UK government ... does not see what value there would be from setting such an arrangement(Respondent #1, UK Department of Health)

We are not a good option for the UK (Respondent #12, India industry association)

When asked about what policy changes would be needed to enable this relationship and what (if any) role the governments on each country should play, there was widespread disagreement amongst the participants. Although most of the UK participants showed some concern regarding the level at which changes should take place, they all believed whatever changes were required simply would not happen.

It is never going to happen ... It could not happen politically. It may make logical sense, but it would never happen in the real world (Respondent #6, UK medical tourism facilitator)

A very sophisticated bilateral agreement would need to be produced. But the NHS is facing bigger problems at the moment, and no government would devote resources to achieving that(Respondent #7, UK think tank)

The Indian participants disagreed on whether the Indian government should be actively involved in facilitating medical tourism. The majority of the respondents, including those from Indian healthcare establishments, believed that the government should be involved, whereas the respondents from the Ministry of Health and academic and industry associations did not.

The government of India should focus on medical value travel as a source of foreign exchange. It can also help improve standards, skills, enable international accreditation and in the process help domestic patients (Respondent #18, Indian healthcare provider)

The official position can never be to focus on medical tourism as this would be perceived as coming at the cost of public health (Respondent #21, Indian Ministry of Health)

Overall, although there was some indication that movement toward a bilateral agreement on medical tourism may make some sense on both sides, the political 'cost' involved was likely to be the major impediment to any movement in practice and in the foreseeable future.

4. DISCUSSION

Of the key themes identified, there was agreement between the stakeholders on the importance of regional travel and proximity, quality, regulation, litigation, continuity of care and the low prospects for a bilateral relationship between India and the UK on medical tourism. However, there were disagreements on the importance of the Diaspora in medical tourism, whether countries like India should concentrate on the local population before offering services to foreign patients or if they should focus on this industry to raise foreign exchange, the status of the three hour 'rule', and what policy changes would be needed for such a relationship to prosper (including the role of government).

These results are subject to two important caveats. First, two different interviewers were involved (one in the UK and one in India), and although the same interview instrument was used there may have been differences in the way questions were asked or answers recorded. Second, the stakeholders approached in India were more willing to participate in the interviews than those approached in the UK. This means that the opinion of those who were not willing to participate from the UK was not represented.

This research supports some of the concerns found in the literature, including quality of care [14], litigation [13, 25], continuity of care [15, 26] and the potentially harmful impact medical tourism may have on the local population [19, 21, 27]. However, as highlighted above, not all stakeholders were in support of these arguments. In addition, this research has shed some light on other issues, such as the uncertainty surrounding three hour 'rule', the importance of the Diaspora and has provided some indication on the amount of foreign patients that Indian private hospitals provide medical treatment to. Although the UK and India were selected as a case study in this research, this is not an exclusive example where this type of trade could take place. The UK has cultural links with many other countries where English is widely spoken that offer medical tourism services, and India attracts medical tourists from across the world.

Nevertheless, when policy makers have to make decisions on whether to engage in this type of trade, this takes place largely in a data vacuum, given that there is no data on the amount of medical tourism that goes on, and what impacts it has on the exporting and importing health system. As this research shows, opinion amongst stakeholders in this field is often divided. These issues cannot be resolved until more and better quality data becomes available.

More research is needed to determine the flows of medical tourism, including the total numbers, what countries they come from and where they get treated. Additionally, it is essential that there is more evidence on the impact medical tourism is having, both on importing and exporting countries, as at present, it is impossible to know whether the concerns outlined in the literature and in these interviews are likely to be realised. Finally, it is of particular importance to examine the effect different types of trade relationships might have. In this respect, whilst the literature review previously carried out by authors suggested that a bi-lateral relationship would be best-placed to address the concerns raised regarding medical tourism, there was scepticism from the analysis provided in this paper based on the over-riding feeling that the political 'cost' involved was likely to be the major impediment. This makes the need for better evidence even more acute, as much of the current policy process could well be based on entrenched ideological positions, rather than secure evidence of impact.

This research can also provide some policy recommendations. The literature suggests that medical tourism can cut costs, improve waiting lists (from the perspective of importing countries) and generate revenue and improve the national healthcare (from the perspective of exporting countries). Given these, policymakers should consider medical tourism as an option to improve their health services. This is particularly important in the current climate where countries are facing budget cuts spurned on by the financial crisis. When evaluating the different options available, policymakers should consider engaging in a bi-lateral relationship, as it has the most potential to ensure the interests of the patients on both countries are prioritised. This could also be addressed through the ongoing EU-India trade negotiations, which also cover services.

In conclusion, there are a wide variety of barriers affecting medical tourism, including restrictions on travel by governments and concerns about quality, regulations and litigation. In order to increase trade, these barriers need to be addressed. The results from the literature review reported elsewhere [Smith R, Martínez Álvarez M, Chanda R: Medical Tourism: a review of the literature and analysis of a role for bi-lateral trade, Submitted] suggests that a bi-lateral relationship is best-placed to tackle these barriers. However, results from the analysis reported here suggest that it is the political sensibility that is the major barrier to be overcome. In this respect, it is worth remembering that policy decisions affecting this type of trade in health services are largely taking place in a data vacuum. It is therefore imperative for more studies and surveys to be carried out and for more empirical data to be collected, in particular data on the size of the medical tourism market, pricing of services, regional breakdown of markets, and the impact of different forms of trade agreement.

ACKNOWLEDGEMENTS

Support was provided by the UK-India Education and Research Initiative, funded by the Department for Innovation, Universities and Skills (DIUS), the FCO, British Council, Department of Science and Technology, Government of India, The Scottish Government, Department of Learning, Northern Ireland, Welsh Assembly, GSK, BP, SHELL and BAE for the benefit of the UK and Indian Higher Education Sector. The views expressed are not necessarily those of the funding bodies.

AUTHORS' CONTRIBUTIONS

MMA contributed to the development of the survey instrument, carried out some of the interviews and drafted the initial paper and agreed on final version

RC conceptualised the work, contributed to the development of the survey instrument, carried out some of the interviews and reviewed, commented and agreed on final version

RS conceptualised the work, contributed to the development of the survey instrument, reviewed, commented and agreed on final version

REFERENCES

1. Smith RD, Chanda R, Tangcharoensathien V: Trade in health-related services. Lancet. 2009, 373 (9663): 593-601.

2. Blouin C, Drager N, Smith R: International Trade in Health Services and the GATS: Current Issues and Debates. 2005, Washington, D.C.: World Bank

3. Blouin C, Drager N, Smith R: Building a national health strategy for trade: A guide for policy making. Montreal: McGill University Press,

4. Lunt N, Carrera P: Medical tourism: assessing the evidence on treatment abroad. Maturitas. 2010, 66 (1): 27-32.

5. Smith RD, Lee K, Drager N: Trade and health: an agenda for action. Lancet. 2009, 373 (9665): 768-773.

6. Mattoo A, Rathindran R: How health insurance inhibits trade in health care. Health Aff (Millwood). 2006, 25 (2): 358-368.

7. Carrera P, Bridges J: Globalization and healthcare: understanding health and medical tourism. Expert Review Pharmacoeconomics Outcomes Research. 2006, 6 (4): 447-454.

8. Crooks VA, Kingsbury P, Snyder J, Johnston R: What is known about the patient's experience of medical tourism? A scoping review. Bmc Health Serv Res. 2010, 10: 266

9. Confederation of Indian Industries and McKinsey & Co: Healthcare in India: The road ahead. 2002, New Delhi: CII & McKinsey & Company

10. Deloitte Center for Health Solutions: Medical Tourism: Consumers in search of value. 2008, Washington, D.C.: Deloitte

11. Tourism Research And Marketing: Medical tourism: a global analysis. 2006, Arnhem: ATLAS

12. Canales MT, Kasiske BL, Rosenberg ME: Transplant tourism: Outcomes of United States residents who undergo kidney transplantation overseas. Transplantation. 2006, 82 (12): 1658-1661.

13. Chinai R, Goswami R: Medical visas mark growth of Indian medical tourism. Bull World Health Organ. 2007, 85 (3): 164-165.

14. MacReady N: Developing countries court medical tourists. Lancet. 2007, 369 (9576): 1849-1850.

15. Connell J: Tummy tucks and the Taj Mahal? Medical tourism and the globalization of health care. Tourism Management. Edited by: Woodside A, Martin D. 2008, CAB International, 232-244.

16. Sen Gupta A: Medical tourism in India: winners and losers. Indian J Med Ethics. 2008, 5 (1): 4-5.

17. Arunanondchai J, Fink C: Trade in health services in the ASEAN region. Health Promot Int. 2006, 21 (Suppl 1): 59-66.

18. Hopkins L, Labonte R, Runnels V, Packer C: Medical tourism today: what is the state of existing knowledge?. J Public Health Policy. 2010, 31 (2): 185-198.

19. Johnston R, Crooks VA, Snyder J, Kingsbury P: What is known about the effects of medical tourism in destination and departure countries? A scoping review. Int J Equity Health. 2010, 9: 24

20. Reddy S, Qadeer I: Medical tourism in India: Progress or predicament?. Economic and Political Weekly. 2010, 45 (20): 69-75.

21. Shetty P: Medical tourism booms in India, but at what cost?. Lancet. 2010, 376 (9742): 671-672.

22. Coast J: The appropriate uses of qualitative methods in health economics. Health Econ. 1999, 8 (4): 345-353.

23. Braun V, Clarke V: Using thematic analysis in psychology. Qualitative Research in Psychology. 2006, 3: 77-101

24. Hsieh HF, Shannon SE: Three approaches to qualitative content analysis. Qual Health Res. 2005, 15 (9): 1277-1288.

25. Turner L: "First world health care at third world prices": globalization, bioethics and medical tourism. BioSocieties. 2007, 2: 303-325.

26. Cheung IK, Wilson A: Arthroplasty tourism. Med J Aust. 2007, 187: 11-12.

27. Garud AD: Medical tourism and its impact on our healthcare. Natl Med J India. 2005, 18 (6): 318-319.

Chapter 6

Canadian medical tourism companies that have exited the marketplace: Content analysis of websites used to market transnational medical travel

Leigh Turner

Center for Bioethics, University of Minnesota, Minneapolis, USA (55419), USΛ

ABSTRACT

Background

Medical tourism companies play an important role in promoting transnational medical travel for elective, out-of-pocket medical procedures. Though researchers are paying increasing attention to the global phenomenon of medical tourism, to date websites of medical tourism companies have received limited scrutiny. This article analyzes websites of Canadian medical tourism companies that advertised international healthcare but ultimately exited the marketplace. Using content analysis of company websites as an investigative tool, the article provides a detailed account of medical tourism companies that were based in Canada but no longer send clients to international health care facilities.

Methods

Internet searches, Google Alerts, searches on Google News Canada and ProQuest Newsstand, and searches of an Industry Canada database were used to locate medical tourism companies located in Canada. Once medical tourism companies were identified, the social science research method of content analysis was used to extract relevant information from company websites. Company websites were analyzed to determine: 1) where these businesses were based; 2) the destination countries and medical facilities that they promoted; 3) the health services they advertised; 4) core marketing messages; and 5) whether businesses marketed air travel, hotel accommodations, and holiday excursions in addition to medical procedures.

Results

In total, 25 medical tourism companies that were based in Canada are now defunct. Given that an estimated 18 medical tourism companies and 7 regional, cross-border medical travel facilitators now operate in Canada, it appears that approximately half of all identifiable medical tourism companies in Canada are no longer in business. 13 of the previously operational companies were based in Ontario, 7 were located in British Columbia, 4 were situated in Quebec, and 1 was based in Alberta. 14 companies marketed medical procedures within a single country, 9 businesses marketed health care at 2 or more destination nations, and 2 companies did not specify particular health care destinations. 22 companies operated as "generalist" businesses marketing many different types of medical procedures. 3 medical tourism companies marketed "specialist" services restricted to dental procedures or organ transplants. In general, medical tourism companies marketed health services on the basis of access to affordable, timely, and high-quality care. 16 businesses offered to make travel arrangements, 20 companies offered to book hotel reservations, and 17 medical tourism companies advertised holiday excursions.

Conclusions

This article provides a detailed empirical analysis of websites of medical tourism companies that were based in Canada but exited the marketplace and are now inoperative. The article identifies where these companies were located in Canada, what countries and health care facilities they selected as destination sites, the health services they advertised, how they marketed themselves in a competitive environment, and what travel-related services they promoted in addition to marketing health care. The paper reveals a fluid marketplace, with many medical tourism companies exiting this industry. In addition, by disclosing identities of companies, providing their websites, archiving these websites or print copies of websites for future studies, and analyzing content of medical tourism company websites, the article can serve as a useful resource for future studies. Citizens, health policy-makers, clinicians, and researchers can all benefit from increased insight into Canada's medical tourism industry.

KEYWORDS

medical tourism; Canada; cross-border; healthcare; globalization; websites

1. INTRODUCTION

The phrase "medical tourism" is often used to describe individuals travelling for health care and paying out-of-pocket for elective medical procedures [1, 2, 3, 4, 5, 6, 7, 8]. Medical tourism can involve travel within the borders of the country inhabited by a particular patient [9]. However, the term usually refers to transnational travel organized for the purpose of receiving medical care [10]. "Medical tourism" sometimes refers to regional health-related travel across borders dividing one country from another. For example, Canadians seeking health care and travelling from Ontario or Québec to Michigan or Vermont are occasionally described as medical tourists [11]. More commonly, the phrase is used to describe long-distance travel such as when a resident of Canada travels to India for medical care [12]. Individuals requiring emergency treatment seek immediate access to local hospitals and clinics. In contrast, medical tourism generally involves travel for elective, non-urgent medical interventions such as hip and knee replacements, dental procedures, and spinal surgery [13]. Patients might desire prompt access to these interventions but the treatments do not fall into the category of emergency care for life-threatening health conditions. Travel to spas, resorts, hot springs, and healing retreats is often characterized as "health tourism", "wellness tourism", and "spa tourism" [14, 15]. The phrase "medical tourism" is usually reserved for trips involving diagnostic tests and medical procedures falling within the scope of biomedicine. This distinction is not absolute. Many individuals combine different healing modalities and when travelling to international medical facilities seek both biomedical procedures and local healing traditions.

Travel for the purpose of obtaining medical care occurs for many different reasons and must be examined within the context of individual patient circumstances and large social-structural forces or political economies shaping access to health care [16]. For example, some uninsured residents of the U.S. travel to India and Mexico for medical care. Such individuals are often labeled medical tourists but other commentators, noting these travellers' lack of health insurance and inability to gain access to affordable care at local medical institutions, describe them as "medical refugees" or "medical exiles" [17].

Distinguishing among different types of health-related travel, researchers identify various kinds of medical tourists. Individuals travelling for kidney and liver transplants, for example, are sometimes given the label "transplant tourists" [18]. Patients seeking stem cell injections at clinics based in China, India, Mexico, Russia, and elsewhere are often called "stem cell tourists" [19]. The term "reproductive tourist" is used to describe individuals travelling for in vitro fertilization, other types of reproductive technologies, and commercial surrogacy [20]. Some scholars even use the term "suicide tourist" to describe individuals who travel to Switzerland for assistance in suicide [21]. Several of these labels are rather jarring and researchers disagree about the adequacy of various terms used to describe different types of cross-border medical travel. The primary objection to the medical tourism label is the linkage of medical procedures with the holiday-going, pleasure-seeking, leisure activities, and relaxation that the word "tourism" signifies to many individuals [22]. Though disagreement persists about what terms are best regarded as marketing labels and which phrases ought to be used by researchers, the phrase "medical tourism" is now routinely used in newspaper articles, policy documents, guidelines for patients, and peer-reviewed publications.

The subject of medical tourism attracts the attention of researchers from numerous disciplines. Scholars from anthropology, bioethics, cultural studies, law, public health, sociology, women's studies, and other fields have all made important contributions to the subject. Health law scholars provide insight into legal issues associated with cross-border medical travel [23]. Economists examine benefits and harms associated with increased international trade in health services [24, 25]. Bioethicists identify ethical issues and policy concerns they associate with medical tourism and the emergence of an inadequately regulated global marketplace in health services [26, 27]. In particular, they raise questions about quality of care and patient safety, continuity-of-care, commercialization of health care, public health considerations, health equity, and commodification of bodies of individuals selling organs or engaging in commercial surrogacy. Though previous research examining the phenomenon of medical tourism displays interest in how medical travel is situated in particular social, political, and economic contexts, many scholars note the need for more richly detailed, empirically-informed studies that address such topics as why patients travel for care, what types of procedures they undergo, where they travel, what quality of care they receive when they travel to international medical destinations, and what happens if they experience complications after receiving care abroad and then return to their local communities [28, 29].

2. BACKGROUND

The study of medical tourism companies, because of their intermediary role in facilitating medical travel, is an important component of the turn toward more empirically-informed and socially situated studies of medical tourism. To date, researchers studying medical tourism companies have paid particular attention to medical tourism companies based in the United States [30, 31, 32]. Journalists mention the existence of various medical tourism companies in Canada, and several articles in peer-reviewed publications identify different medical tourism companies in Canada, but there is limited scholarly research examining the medical tourism industry in Canada [33, 34, 35]. This article responds to a gap in scholarship and seeks to promote increased understanding of medical tourism in Canada. The text provides a detailed account of where defunct Canadian medical tourism companies were based, the destination sites they marketed to clients, the health services they promoted, and what they advertised in addition to medical procedures. This topic was addressed by compiling, between 2006 and 2011, a database of medical tourism companies based in

Canada. The database contains companies marketing health services at international health care facilities and functioning as medical tourism agencies. It does not include websites that simply provide information about medical tourism. Once a database of medical tourism companies was established, content analysis of company websites was used to: 1) identify and document where in Canada particular medical tourism companies are based; 2) identify and summarize the destination countries to which these businesses indicate they send clients; 3) identify and record medical procedures, tests, and treatments marketed by these businesses; 4) identify and summarize the core marketing message of each of these companies; and 5) address whether medical tourism companies are involved in booking flights, arranging accommodations at hotels and resorts, and offering holidays at resorts, side trips, holiday excursions, visits to local landmarks and other activities often associated with the word, "tourism". The social science method of content analysis was used to address these five topics by analyzing websites of medical tourism companies. After content analysis was completed, medical tourism companies that currently market health care abroad were distinguished from defunct businesses. Medical tourism companies remaining in business are examined elsewhere. Companies that have ceased operations are the focus of this article. Unlike previous contributions to the study of transnational medical travel, the purpose of this article is to identify and analyze medical tourism companies that have exited the marketplace for transnational medical travel. In addition to its analytic function, the paper has historical value by capturing a turbulent period in the medical tourism marketplace and examining those Canadian businesses that are no longer involved in competing for clients. Medical tourism is often described as an industry with a rapidly expanding client base. This paper offers a more complicated account of the market for medical tourism by revealing that of all the medical tourism companies in Canada that could be identified over the last five years, approximately half of them are no longer in business.

3. METHODS

3.1 Development of a database of medical tourism companies located in Canada

Before analyzing websites of individual medical tourism companies it was first necessary to develop a database of all medical tourism companies with head offices or affiliate offices in Canada. Several methods were used to build this database. From 2006 to 2011, repeated Internet searches using Google Search Engine played a key role in locating medical tourism companies situated in Canada. Many businesses came into existence over the course of this study and were not identifiable during initial searches for medical tourism companies in Canada. Such phrases as "medical tourism company Canada", "medical tourism agency Canada", and "medical tourist Canada" were used as search terms. Searches for newspaper articles addressing such topics as "medical tourism", "medical tourism Canada", and "cross-border healthcare Canada" provided additional leads [36, 37, 38, 39,40]. Newspaper articles were located using repeated searches on Google News Canada and ProQuest Newsstand. Newspaper reports of medical tourism companies situated in British Columbia, Alberta, Manitoba, Ontario, and Quebec played an important role in tracking the emergence of medical tourism companies across Canada. Google alerts for such terms as "medical tourism Canada", "stem cell tourism Canada", "transplant tourism Canada", and "global health care Canada" also generated extensive news reportage of Canadian medical tourism companies and Canadian patients travelling abroad for care. In addition, searches of an Industry Canada website that lists federally incorporated businesses in Canada yielded three company names. However, despite repeated efforts to locate these businesses it was not possible to find company websites, news media coverage concerning these businesses, or other signs indicating that these companies successfully entered the marketplace for promoting transnational medical travel. These companies are therefore not included in the final tally of twenty-five defunct companies with identifiable websites. Once company websites were identified they were bookmarked and added to the database. To facilitate content analysis, company websites were printed in their entirety. Content analysis was performed by analyzing printed versions of company websites and by periodically revisiting the Internet websites of medical tourism companies. Electronically archived websites or printouts of all identified medical tourism companies are available for review. Readers interested in assessing the content analysis provided in this article are welcome to access the primary data of the company websites.

3.4 Comparing the database to other lists of medical tourism companies in Canada

Internet searches resulted in the identification of several sources providing partial lists of medical tourism companies found in Canada. For example, in 2006, *Maclean's*, a popular news magazine in Canada, published an article that identified eight medical tourism companies in Canada [41]. Another list identifying five medical tourism businesses was found on a website promoting private health care alternatives and out-of-country health care options for Canadians [42]. A third source listed eight Canadian medical tourism companies within a longer record of medical tourism companies based in countries around the world [43]. Locating these lists permitted comparison of my database with other summaries of medical tourism companies based in Canada.

3.5 Content analysis of websites of medical tourism companies

After developing the database content analysis was used to study and analyze websites of all identified medical tourism companies based in Canada. Company websites were analyzed using pre-determined analytic categories; content was extracted for each category. The social science method of content analysis, as the phrase suggests, involves studying the content of such media as magazines, newspapers, films, television programs, and interview transcripts [44]. Content can be analyzed for general themes and concepts, images, particular statements, and other features. For the purpose of this study, medical tourism company websites were searched for specific pieces of information or website content. There were five main components to content analysis.

3.6 Location of medical tourism companies

First, I identified the city or town and province that medical tourism company websites identified as their business address within Canada. Identifying the location of medical tourism companies typically involved finding and then selecting such website tabs as "Contact us", "Contact information", or "Address" and recording information indicating where companies are located. Gathering this information resulted in a detailed record of where in Canada now-defunct medical tourism companies were based. In particular, it helped establish where they clustered. Though medical tourism companies in some respects are based "on" the internet, the physical location of companies matters for numerous reasons. For example, companies are governed by the legislation and regulations of the specific regions in which they are located. Within Canada, companies must operate according to both federal and provincial legislation; the province in which companies are situated informs which legal standards can be applied to them.

3.7 Destination countries and health care facilities

Second, I recorded the destination countries and/or destination health care facilities that now-inoperative medical tourism companies once marketed to their clients. Some medical tourism company websites identified particular hospitals and clinics. Other websites mentioned destination nations or networks of countries but did not list specific medical facilities. Websites were analyzed to develop a better understanding of destinations promoted by Canadian medical tourism companies.

3.8 Types of advertised medical procedures and medical specialties

Third, to provide insight into the types of health care packages medical tourism companies promoted, I recorded medical procedures or specialties marketed by medical tourism companies. Some companies provided lengthy lists of procedures. Other companies identified particular areas of medicine without listing specific procedures. Other businesses mentioned clinical specializations such as cardiology and gastroenterology or listed illnesses rather than focusing upon distinct medical procedures. This topic was explored to better understand what types of tests and treatments medical tourism companies marketed to prospective clients.

3.9 Core marketing messages of medical tourism companies

Fourth, websites of medical tourism companies were reviewed for statements that best encapsulated each company's core marketing message. These statements were typically located on the home page of company websites or in such sections as "About us", "What We Do", or "Mission statement". Core marketing messages were recorded and then summarized in brief point form. This topic was studied to develop a detailed account of the market niche now-defunct Canadian medical tourism companies attempted to occupy while they remained operational.

3.10 "Tourism" component of medical tourism

Fifth, company websites were reviewed for information about involvement of these businesses in booking air and/or ground transport, reserving hotel accommodations, and organizing tours, side trips, and/or other holiday excursions. This topic was studied in an attempt to determine whether holidays, side trips and other tourism-related activities were marketed in addition to health care packages.

In most instances, websites provided detailed information for each study question or category of content analysis. Data was recorded in tabular form. Where websites did not provide pertinent information, "NA" ("Not Addressed" in abbreviated form) was noted in tables.

The author of this article performed content analysis. The paper was fact-check three times by the author and feedback was solicited from two senior colleagues. All websites are disclosed and electronically archived. Readers interested in assessing the quality of content analysis have access to the primary data.

3.11 Distinguishing functioning from inoperative medical tourism companies

Following content analysis of all identified medical tourism company websites, in February and March 2011, internet searches, searches of an

Industry Canada database [45] identifying the status of federally incorporated businesses in Canada, phone calls, and emails were used to distinguish active medical tourism businesses from medical tourism companies with expired websites, non-functioning email accounts, and discontinued phone numbers. Email and phone queries were used to determine whether businesses remained in operation. According to research ethics guidelines in both Canada, where this research project was initiated, and the United States, where it was completed, contacting a company to see if it remains in business does not fall within the scope of research requiring research ethics board review because there was no attempt to ask company owners or employees about their role within these companies, explore their attitudes toward medical tourism, investigate their workplace activities, or otherwise treat them as research subjects [46].

4. RESULTS

Using Internet searches, searches for newspaper articles mentioning medical tourism companies, and Google Alerts, it was possible to locate a total of twenty-five Canada-based medical tourism companies that have ceased sending their clients to hospitals and clinics located outside Canada. In contrast, operational medical tourism companies include an estimated 18 businesses marketing medical travel to such countries as Costa Rica, India, and Thailand and 7 businesses marketing regional, cross-border health services available in the United States as well as travel to private clinics within Canada. Several additional businesses are not comprehensive medical tourism companies but market bariatric surgery procedures performed in facilities based outside Canada as well as so-called "Liberation therapy" performed in India. Excluding these latter "boutique" businesses, approximately half of all medical tourism companies in Canada remain operational and the other half has disappeared from the marketplace.

Companies were labeled inoperative if websites were deactivated and phone lines were disconnected, company representatives reported that they were no longer accepting clients and had halted operations, they had lost their status as federally incorporated companies, or seven phone calls and/or emails failed to generate any response. It is possible that businesses that at present are not going concerns could at some point build a client base and begin sending customers to international medical facilities. If this occurs, in future analyses these companies will be reclassified as functioning medical tourism companies. My research provides a "snapshot" of Canada's medical tourism "landscape" at a particular moment; it is important to appreciate that the medical tourism industry in Canada is in flux. It is reasonable to assume that this industry is going to change over time as some companies emerge, other businesses disappear, and other companies transform themselves in response to client demand and perceived marketing opportunities.

Of the inoperative companies that I was able to locate, twenty-five had active, functioning websites at some point between 2006 and 2011. Additional file 1 identifies these businesses, provides website links, and contains Webcite references. Three additional companies, St. Luke Medical Tourism Center of North America Inc., Medical Tourism China, and International Medical Travel Corporation, were identified in the Government of Canada's database of businesses federally incorporated in Canada but do not appear to have had websites, issued press releases, or attracted any news media coverage. It is unclear whether these companies ever had clients even though they were incorporated businesses presumably established for the purpose of marketing medical tourism. These three businesses are noted but they are not included in the total tally of twenty-five defunct medical tourism companies. With no websites to analyze, and no indications that they had ever sent clients to international destinations, there was no basis for including them in the content analysis process.

4.1 Locations of medical tourism companies

Of the twenty-five companies that established websites and were studied using the method of content analysis, thirteen were located in Ontario, seven were based in British Columbia, four were situated in Quebec, and one was located in Alberta. Websites of some Canadian medical tourism companies identify affiliate offices or company representatives located outside Canada. Where such information was provided I recorded where company representatives based outside Canada were situated.

Table 1 lists company names, identifies where in Canada these businesses were located, and documents affiliates and representatives in those instances where companies had offices or agents situated outside Canada.

Table 1. Locations of Medical Tourism Companies in Canada

Company	City & Province
Axiom Health Solutions	Canadian Office: Oakville, Ontario; U.S. Office: Kensigton [sic], Maryland
Canadian Healthcare International (CHI)	Markham, Ontario
CubaMedicare	Oakville, Ontario
EcuMedical Resources International Ltd.	Windsor, Ontario
First Choice Medical Tourism	Canadian Office: Calgary, Alberta; with contact person in Luzon, Philippines
Health Trips	Ottawa, Ontario
Health Vacations, Inc	Ottawa, Ontario
International Medical Network	Oakville, Ontario
JD Healthcare	Nanaimo, British Columbia
LAM International (L.A.M. Logistic.Assistance.Medical International)	Canadian Office: Toronto, Ontario; and Colombia
MedAsia	Montreal, Quebec
MedExpress Tourism	Montreal, Quebec
Medi-Pro Medical Management	Windsor, Ontario
MedSolution	Vancouver, British Columbia
Medtourlink	Vancouver, British Columbia
Reach Health Services & Outsourcing	Canadian Office: Vancouver, British Columbia; Representative in Chennai, India
Recover Discover Healthcare	Canadian Office: Vancouver British, Columbia; U.S. Office: Austin, Texas; India Office: Noida, India

Royal Med Services	Mississauga (Toronto), Ontario
Speedy Surgery Global Healthcare	Quebec City, Quebec
Star Hospitals	Canadian marketing office in Toronto, call centers and operations in Chennai, India
Sun Medical Group	Vancouver, British Columbia
The InciDental Tourist	Ottawa, Ontario
Tooth Tourism	Surrey, British Columbia
Unbelievable India	Dollard-Des-Ormeaux, Quebec
Victus Global Healthcare	Main Office: Toronto, Ontario; Additional office: Edmonton, Alberta

4.2 Destination countries and health care facilities

Of the twenty-five companies, fourteen marketed medical procedures at just one medical facility or at several medical facilities within one country. Three companies marketed health care in two destination nations, two companies advertised procedures in three countries, two companies promoted health services in four countries, one company marketed procedures in five countries, and one of the dental tourism companies marketed dental procedures in seven countries. Two companies did not provide details concerning where they sent their clients.

The twenty-five medical tourism companies marketed health services in twenty-one different nations. Twelve companies marketed medical travel to India; four listed Thailand as a destination site; three companies listed Singapore; Canada, Cuba, the United States, the Philippines, Costa Rica, El Salvador, China, Malaysia, and Mexico were all twice listed as destination sites; and Morocco, Russia, Hungary, Colombia, Taiwan, France, Turkey, Dominican Republic, and Panama were all listed once. Listing of countries as destinations does not mean that any residents of Canada ever selected these particular destinations as sites for medical care. However, documentation of advertised health care destinations provides insight into the partnerships or networks medical tourism companies choose to market to their prospective client base. While they do not provide information about actual flows of patients across national borders, they reveal how medical tourism companies that were based in Canada promoted health facilities in other nations. Table 2 identifies locations of destination facilities marketed on company websites.

4.3 Types of advertised medical procedures

Twenty-two companies marketed many different kinds of medical procedures and can be characterized as "generalist" medical tourism companies. They provide lengthy lists of different types of medical procedures. Three businesses marketed specialized services and restricted themselves to a niche position within the medical tourism industry. Of these latter businesses, two companies limited themselves to marketing dental procedures and one company advertised organ transplants in Colombia. Table 3 summarizes health services marketed by medical tourism companies.

Table 2. Medical Tourism Companies and Destination Facilities

Company	Destination Facilities
Axiom Health Solutions	Apollo Hospitals, Wockhardt Hospitals, Escorts Heart Institute and Research Centre, Max Healthcare, Fortis Healthcare, Sahaj Dental Clinic, all in India
Canadian Healthcare International (CHI)	Health care facilities in Canada (particular institutions are not identified)
CubaMedicare	Medical facilities in Havana, Varadero, and Holguin, all in Cuba
EcuMedical Resources International	Health care facilities in the United States
First Choice Medical Tourism	St. Luke's Medical Center, Quezon City, Philippines; Estetico Manila, Manila, Philippines
Health Trips	German Collaborative Private International Hospital, India
Health Vacations, Inc	St. Petersburg, Russia
International Medical Network	Hospitals and clinics in India, Hungary, Cuba, and Costa Rica
JD Healthcare	Website claims company has partnerships with more than 30 hospitals (particular institutions are not specified)
LAM International (Logistic Assistance. Medical International)	Transplant facilities in Colombia
MedAsia	Min-Sheng General Hospital, Taiwan
MedExpress Tourism	Health care facilities in Morocco
Medi-Pro Medical Management	Laser Spine Institute, Tampa, Florida; Crittenton Hospital, Rochester, MI, Internal Medicine, Mount Clemens, MI; additional facilities in the U.S.
MedSolution	India, France, Turkey, El Salvador; website indicates plan to include Costa Rica, Malaysia, Mexico, Poland, and South Africa

Medtourlink	Singapore, Philippines, India, China, Canada
Reach Health Services & Outsourcing	Apollo Hospitals, Escorts Heart Institute and Research Centre, Rajan Eye Care, JGHR Dental, all in India
Recover Discover Healthcare	Escorts Heart Institute & Research Centre, Delhi; Fortis Hospital, Noida; Fortis Flt. Lt Rajan Dhall Hospital, Delhi; Fortis Hospital, Mohali, Chandigarh; Fortis-Wockhardt Hospital, Bangalore; Fortis-Wockhardt Hospital, Mumbai, India; Gleneagles Medical Center, Penang, Malaysia
Royal Med Services	Medical facilities in India
Speedy Surgery Global Healthcare	Medical facilities in India and Thailand
Star Hospitals	India, Singapore, Thailand
Sun Medical Group	Malaysia, Singapore, Thailand
The InciDental Tourist	Nanjing,China; Merida, Mexico
Tooth Tourism	Costa Rica, Dominican Republic, El Salvador, India, Mexico, Panama, Bangkok, Thailand
Unbelievable India	Medical facilities in India
Victus Global Healthcare	JCI-accredited medical facilities around the world (particular institutions are not specified)

Table 3. Marketed Procedures

Company	Health Services Marketed
Axiom Health Solutions	Interventional cardiology, cardiothoracic surgery, cosmetic and plastic surgery, hand and micro surgery, medical & surgical gastroenterology, neurology & neurosurgery, spine surgery, surgical oncology, orthopedic surgery, vascular surgery, brain surgery, bone and joint surgery, eye surgery, minimal access surgery, obesity surgery, maxillofacial surgery, thoracic surgery, orthopedic surgery, dental care and dental surgery, transplant surgery, and other treatments
Canadian Healthcare International (CHI)	Specific procedures are not identified though total hip replacement and arthroscopic rotator cuff repair are used to compare cost of care in Canadian facilities versus institutions in other countries
CubaMedicare	Drug and alcohol rehabilitation, pigmentary retinosis, neurological rehabilitation, vitiligo, psoriasis, and alopecia, cosmetic surgery, odontology, dental implants, orthognatic surgery; alternative therapies, stroke rehabilitation, spinal cord injury rehabilitation, brain injury rehabilitation, physical, occupational, and speech therapy, aquatic therapy, physical therapy, anorexia nervosa treatments
EcuMedical Resources International Ltd.	cancer, cardiology, cosmetic surgery, day surgery, diagnostic imaging, hip surgery, knee surgery, minimally invasive surgery, neck and spine surgery, neurology, orthopaedics, second opinions, urology, VIP services
First Choice Medical Tourism	General medical surgery, dental surgery, cosmetic surgery, optical surgery, spa retreats
Health Trips	knee replacement, hip replacement, shoulder replacement, spinal surgery, correction of deformities, beating heart surgery, cosmetic surgery
Health Vacations, Inc	MRI outsourcing, endocrinology surgery, orthopedic packages, ophthalmology packages, spinal surgery, diagnosis and treatment of allergies, preventive health check ups, executive health check ups, preventive heart checkup, whole body check up, and additional treatments, access to specialists in cardiology, internal medicine, neurology, obstetrics/gynecology, orthopedic surgery, pediatrics, urology, endocrinology, rheumatology
International Medical Network	Cardiovascular, orthopedic, neurosurgery, female reproductive system, gastro-intestinal, cosmetic, dental, IVF

JD Healthcare	Assisted reproduction, plastic surgery, bone and joint surgeries, cancer treatment, cardiovascular surgery, eye surgery, general surgery, cerebrovascular surgery, dental surgery, dermatology, and additional treatments
LAM International (Logistic. Assistance. Medical International)	Heart transplants, liver transplants, lung transplants, kidney transplants
MedAsia	Bariatric surgery, cardiovascular surgery including minimally invasive cardiovascular surgery, PTCA and stent, and cardiac catheterization, orthopedic surgery including total joint replacement, spine surgery, and arthroscopy, other treatments
MedExpress Tourism	Dental implants, weight loss, cosmetic surgery, orthopedics, urology, cardiology, eye corrective (Lasik), cancer, general surgery
Medi-Pro Medical Management	MRI scans, CT scans, body imaging, X-rays, arthroscopy, bariatric surgery, cardiology, endocrinology, internal medicine, neurology
MedSolution	Cardiac care, chemical dependency programs, cosmetic & plastic surgery, dental care, gastric surgery, heart surgery, infertility treatments, orthopedic surgery, urogenital surgery
Medtourlink	Hysterectomy, cataract, Lasik eye care procedures, breast lift, tummy tuck, facelift, rhinoplasty, dental implants, dentures, crowns and aesthetic dentistry, MRI, X-ray, PET, CT diagnostic scans, laser procedures
Reach Health Services & Outsourcing	Cardiac surgery and cardiology, minimal access surgery and urology, orthopedics, ophthalmology, gastroenterology, obstetric and gynecology, oncology, cosmetic/plastic surgery, preventive health checks, diagnostic services
Recover Discover Healthcare	Knee replacement surgery, hip replacement surgery, hip resurfacing, adult & pediatric cardiac care, oncology treatments, bariatric care, laparoscopic gastric band, gastric bypass surgery, sleeve gastrectomy, neurosurgeries, spinal care, infertility management including IVF, cosmetic care, additional surgeries

Royal Med Services	Hip replacements, other joint replacements, breast augmentation, liposuction, and other procedures
Speedy Surgery Global Healthcare	Allergy, cardiology, dentistry, diabetes and endocrinology, ear, nose and throat diseases, eye laser refraction, oncology, ophthalmology, orthopedic, plastic surgery, preventive health check, radiology and imaging services, and additional procedures
Star Hospitals	cardiology, cardiothoracic surgery, orthopedic, neurology and neurosurgery, eye care, cosmetic surgery, dentistry, comprehensive and preventive health checks, weight loss, obstetrics and gynecology, IVF, additional treatments, yoga, Ayurvedic consultation, stress management, weight reduction, spine and joint care
Sun Medical Group	Dental surgery, eye care, cardiac interventions, orthopaedic surgery, cosmetic surgery
The InciDental Tourist	bridgework, partial and full crowns, laser tooth bleaching, implants, inlays, onlays, veneers, root canals, extractions, and other dental procedures
Tooth Tourism	Dental bonding and contouring, dental bridge, dental crowns, dental fillings, dental implants, dental veneers, dentures, gingivectomy, root canal, teeth whitening
Unbelievable India	Plastic surgery, correction of congenital malformations, reconstructive surgery, orthopedics, ophthalmology, hip replacements, Ayurvedic medicine, dental procedures, other treatments
Victus Global Healthcare	Heart bypass, heart valve replacement, hip replacement, hysterectomy, knee replacement, spinal fusion, additional treatments

4.4 Core marketing messages of medical tourism companies

Though there were variations in core marketing messages of medical tourism companies, most businesses emphasized affordability of care, timely access to medical care, and high-quality care. In total, of twenty-five businesses, twenty companies marketed access to affordable care, seventeen businesses advertized timely access to care, and twenty-one websites emphasized high-quality care. The few companies that did not mention all three features offered a subset of these offerings.

Table 4 provides summaries of medical tourism companies' core marketing messages. Additional file2 provides summaries of core marketing messages in addition to extended excerpts from company websites. These lengthy excerpts are intended to demonstrate the empirical basis for identifying and summarizing core marketing messages.

4.5 "Tourism" component of medical tourism

Core marketing messages rarely emphasized the possibility of patients undergoing medical procedures while also enjoying holiday excursions. However, activities commonly associated with the concept of tourism were marketed by more than half of the companies. Sixteen medical tourism companies offered to arrange air travel and/or organize ground transport. Twenty companies marketed the service of booking hotel reservations. Seventeen businesses advertised tours, side trips, and other holiday excursions in addition to medical care. This latter finding will not resolve disagreements about whether "medical tourism" is a suitable term for academic analysis. However, it does reveal that many medical tourism companies in Canada market both medical care and leisure or "holiday" activities commonly associated with tourism. Table 5 identifies whether medical tourism companies booked travel arrangements, arranged hotel accommodations, and offered holiday excursions, side trips, visits to local attractions, and other "tourist-like" excursions.

Table 4. Summary of Core Marketing Messages

Company	Core Marketing Message
Axiom Health Solutions	access to affordable, timely, and high-quality care
Canadian Healthcare International (CHI)	access to affordable and high quality health care in Canada
CubaMedicare	access to affordable, timely, and high quality health care as well as vacation experience
EcuMedical Resources International Ltd.	access to timely and high-quality health care
First Choice Medical Tourism	access to affordable, timely, and high quality health care as well as holiday experience
Health Trips	access to affordable and high-quality care as well as vacation experience
Health Vacations, Inc.	access to affordable, timely, and high-quality care
International Medical Network	access to affordable, timely and high-quality care
JD Healthcare	access to affordable, timely and high-quality health care
LAM International (Logistic.Assistance.Medical International)	access to timely organ transplants
MedAsia	access to affordable, timely, and high-quality care
MedExpress Tourism	access to affordable and high-quality care
Medi-Pro Medical Management	access to timely and high-quality care
MedSolution	access to affordable, timely and high-quality care
Medtourlink	access to high-quality care
Reach Health Services & Outsourcing	Access to affordable and timely health care
Recover Discover Healthcare	access to affordable and high- quality care
Royal Med Services	access to high-quality health care
Speedy Surgery Global Healthcare	access to affordable, timely and high-quality health care

Star Hospitals	access to affordable, timely, and high-quality health care
Sun Medical Group	access to affordable and timely care
The InciDental Tourist	access to affordable and high-quality care
Tooth Tourism	access to affordable and high-quality dental care
Unbelievable India	access to affordable, timely, and high-quality care
Victus Global Healthcare	access to affordable and timely health care (messages targeted at employers rather than individuals)

Table 5. Medical Tourism Companies Marketing Travel, Accommodations, and Tourism-related Activities (Y = Yes, N = No, NA = Not Addressed on company website)

Company	Book Travel	Book Hotel	Book Tours
Axiom Health Solutions	Y	Y	Y
Canadian Healthcare International (CHI)	Y	Y	NA
CubaMedicare	Y	Y	Y
EcuMedical Resources International Ltd.	Y	Y	NA
First Choice Medical Tourism	NA	NA	Y
Health Trips	Y	Y	Y
Health Vacations, Inc.	Y	Y	Y
International Medical Network	Y	Y	Y
JD Healthcare	N	NA	Y
LAM International (Logistic.Assistance.Medical International)	NA	Y	Y
MedAsia	Y	Y	Y
MedExpress Tourism	Y	Y	N
Medi-Pro Medical Management	Y	NA	NA
MedSolution	Y	Y	Y
Medtourlink	NA	Y	NA
Reach Health Services & Outsourcing	NA	Y	NA
Recover Discover Healthcare	NA	NA	Y

Royal Med Services	Y	Y	Y
Speedy Surgery Global Healthcare	Y	Y	NA
Star Hospitals	Y	Y	Y
Sun Medical Group	NA	NA	Y
The InciDental Tourist	N	Y	Y
Tooth Tourism	N	Y	Y
Unbelievable India (partnership with Focuz Group)	Y	Y	Y
Victus Global Healthcare	Y	Y	NA

5. DISCUSSION

5.1 Principal Results

5.1.1 Value in studying functioning and inoperative medical tourism companies

To date, there are no published studies that identify by company name, analyze in systematic fashion, disclose in transparent manner, and permanently archive websites of medical tourism companies with head offices or affiliate offices in Canada. Previous articles mention several medical travel facilitators based in Canada but do not provide a comprehensive overview of identifiable medical tourism companies, disclose their websites, analyze their business operations, and clearly distinguish between functioning and defunct companies [47]. This article provides insight into those medical tourism companies in Canada that came into existence and at some point ceased operations between 2006-2011. To some individuals, identifying and analyzing medical tourism companies that have exited the marketplace might seem like an unproductive exercise. However, identifying and analyzing medical tourism companies that have ceased functioning serves two purposes. First, it contributes to the overall analysis of Canada's medical tourism industry rather than presenting just a limited account of businesses successfully involved in advertising health care at international destinations. Analyzing companies that ceased functioning permits insight into where they were based, what health care procedures they marketed, where they proposed sending clients, what part of the marketplace they sought to occupy, and whether they advertised tourism-related activities such as offering side trips and holiday excursions in addition to booking flights and accommodations. Second, and perhaps more importantly, identifying and examining inoperative medical tourism companies challenges the hyperbolic marketing rhetoric surrounding the topic of medical tourism and could play a role in promoting more balanced ethical, social, and economic analysis of the subject [48]. Not all medical tourism companies remain going concerns. Studying websites of medical tourism companies provides no insight into how many Canadians seek care beyond Canada's borders. Nonetheless, perhaps the disappearance of an estimated 50% of all medical tourism companies based in Canada should prompt questions about the overall size of the marketplace within Canada for international medical travel. It is conceivable that the market for medical care abroad is smaller than anticipated and some medical tourism companies fail because their prospective clientele is less numerous than assumed [49]. Another possibility is that of the many medical tourism companies established in Canada, only a limited number of them have successfully developed significant client bases. It is also conceivable that some medical tourism companies were commercially successful but closed for non-financial reasons. This study does not attempt to address these questions but it should prompt critical reflection on the size and evolving shape of Canada's marketplace for medical tourism. Future research will attempt to address the question of why some medical tourism companies exit the marketplace whereas other businesses appear to be successful at attracting clients and expanding their operations.

Key findings from analysis of websites of Canadian medical tourism companies are that twenty-five companies established for the purpose of sending residents of Canada to medical facilities in other nations are no longer operational, most defunct medical tourism companies are "generalist" rather than "specialist" businesses, and many companies offer tourism-related activities in addition to marketing medical procedures.

5.2 Medical tourism company "hotspots"

Of the twenty-five companies with websites, thirteen businesses based in Ontario are now defunct, seven companies in British Columbia ceased operations, four companies in Quebec are closed, and one company in Alberta did not respond to calls and emails and is presumed closed. This study does not address how many individuals travel from Canadian provinces to international medical destinations. However, it is worth noting the concentration of defunct medical travel companies in Ontario, British Columbia, and Québec. These provinces also happen to be the three provinces with the greatest number of medical tourism companies still functioning. Perhaps they should be characterized as "hotspots" for both the establishment and closure of Canadian medical tourism companies. Future research will explore why numerous medical tourism companies have been established in some Canadian provinces whereas other provinces have no businesses involved in the medical tourism industry.

5.3 Choice of marketing messages

There likely are numerous reasons why medical tourism companies based in Canada use particular marketing messages. Most clients of these businesses must pay out-of-pocket for private medical care provided by hospitals and clinics based outside Canada. Though there are some circumstances in which provincial health insurance plans will fund provision of elective medical procedures performed outside Canada, in most instances clients of these businesses must use personal savings to purchase treatment. It is therefore unsurprisingly that affordability of care is emphasized in marketing messages. Access to high-quality care is presumably mentioned as a way to reassure prospective clients that they should feel confident that they will receive professional care when they go abroad for treatment. This message is perhaps promoted as a way of allaying fears that going abroad for inexpensive care might increase risk of exposure to lower-quality care. Finally, the message of gaining access to "timely" health care might have particular resonance for Canadians wait-listed for orthopaedic procedures, ophthalmologic procedures, spinal surgery, bariatric surgery, and other interventions where there are sometimes considerable delays prior to obtaining access to publicly-funded health care at Canadian medical facilities. For example, Speedy Surgery Global Healthcare, one defunct company mentioned in this analysis, and Timely Medical Alternatives, a company that continues sending Canadians to hospitals and clinics in the United States as well as private health care facilities in Canada, both emphasize in their company names provision of prompt access to care. Wanting to avoid providing a reductionist, single-factor account of how medical tourism is marketed by companies based in Canada, it nonetheless seems reasonable to suggest that timely access to care abroad, in contrast to treatment delays for access to some types of care in medical procedures performed in Canada, is one marketing message likely to resonate with Canadians frustrated with wait-times for various elective procedures. In this respect, it is possible that the marketing messages used by medical tourism companies based in Canada might differ somewhat from marketing claims used by medical tourism companies located in countries where the presence of a significant private health sector means that clients with sufficient financial assets can gain prompt access to medical procedures. Comparative studies that contrast marketing messages of medical tourism companies based in different countries should provide insight into regional and national variations in the types of marketing messages used to promote medical tourism.

5.4 Generalist and specialist companies

Another finding that merits mention is that twenty-two of the twenty-five defunct medical tourism companies can be characterized as "generalist" medical tourism companies. This label means that these companies marketed multiple procedures, tests, and treatments rather than confining themselves to dental treatments or some other specialized domain of care. Even companies that offer many different services are at risk of ceasing operations. It is not known whether cessation of operations is linked to a limited client base, an overreliance on one or two health care destinations, concerns about liability, competition from other medical tourism companies, a low ratio of earnings to expenditures, or other factors. Future research will explore why some companies remain operational and other businesses cease functioning.

5.5 Closure of dental tourism companies

Both of the companies that promoted "dental tourism" are no longer operational. While this article does not address why medical tourism companies ceased operations, numerous hypotheses are worth exploring in future studies. Cost savings associated with obtaining dental care abroad might be lower than anticipated once airline tickets, ground transport, hotel accommodations, and other expenses are factored into consideration. Concerns about patient safety and quality of care might deter some Canadians from receiving dental care abroad. It is also conceivable that the practice of dentistry is not well suited to a business model that involves travel to international dental clinics. Many forms of dental care require repeat visits rather than one-time procedures. Having multiple dental procedures at one time could put patients at increased risk of complications. Staying abroad for multiple procedures would presumably increase time away from home and also generate additional expenses. Though this study offers no definitive insights into why both of Canada's dental tourism companies are defunct, it is worth noting that at present there appear to be no functioning companies in Canada that are based exclusively upon a business model of marketing dental care abroad.

5.6 Little marketing of organ transplants performed abroad

Travel abroad for the purpose of purchasing a kidney used in organ transplantation, sometimes described as "transplant tourism", receives considerable attention from both health researchers and journalists. However, of Canadian medical tourism companies no longer in operation, just one business had as its primary message the marketing of organ transplants abroad. One other company marketed organ transplants in addition to other medical procedures. It is conceivable that other medical tourism companies did not market organ transplants abroad because of fears that they might be accused of facilitating "organ brokering" or "organ trafficking"; concerns about negative news media coverage that might harm other aspects of their business; inability to identify facilities able to provide organ transplants; ethical reservations about intentionally or unintentionally participating in organ trafficking; or concerns about legal repercussions. Whatever the reason, with the exception of one business that advertised organ transplants in Colombia and one company that claimed to be able to arrange transplants in India, most now-defunct medical tourism companies in Canada did not, while they remained operational, openly market organ transplants performed in other countries.

5.7 Medical consumerism and restrictions on choice of destination nations

Another noteworthy finding is that of the twenty-five businesses with websites, fourteen of them sent their clients to just one medical facility or destination nation. In this business model, owners of medical tourism companies decide that all their clients will travel to just one location. It is conceivable that this arrangement is not an optimal business model for an industry that in many respects is built upon promoting patient choice and medical consumerism. It is possible that medical tourism companies offer more attractive options to a larger prospective clientele when they market health care facilities at multiple international destinations. Whether or not this claim is credible, it is important to note that even businesses with multiple destinations ceased operations. One dental tourism company, for example, listed seven national destinations on its website. In short, while limiting the number of medical destinations constrains choice and possibly acts as a deterrent to prospective clients, promoting choice in the form of multiple travel destinations does not ensure long-term success in the medical tourism marketplace.

5.8 "Tourism" component of medical tourism

Though some researchers express reservations about the phrase "medical tourism", many now-inoperative medical tourism companies in Canada marketed travel-related services such as booking air and ground transport, organizing hotel accommodations and arranging side trips, and other tourist-like activities. In this aspect of their business models, medical tourism companies resemble travel agencies. These activities provide some insight into how medical care can be connected to activities more typically associated with "traditional" forms of tourism.

5.9 Limited media coverage of defunct medical tourism companies

Whatever causes medical tourism companies to close, it is important to note that in Canada a substantial number of medical tourism companies have exited the marketplace. The closure of approximately half of all medical tourism companies in Canada has not been previously examined and addressed by health researchers and journalists. Whereas medical tourism companies routinely issue press releases and attract news media coverage when they commence operations, these businesses typically do not issue press releases or receive news coverage when they close. Just one business, EcuMedical Resources International, generated news coverage as it ceased functioning [50,51]. In that company's case, several of its clients claimed that they paid EcuMedical for arranging medical procedures in the U.S. and then began receiving bills from the hospitals where they received care. Following a police investigation the two company owners were charged with multiple counts of fraud. In this case, activity that is presumably atypical in the medical tourism industry resulted in considerable local news coverage of a medical tourism company ceasing operations.

5.10 Basis for additional research

One important benefit of studying medical tourism companies by analyzing content of medical tourism company websites is that the exercise can challenge hitherto unexamined assumptions about the industry. More fine-grained analyses that draw upon multiple research methods such as surveys, in-depth interviews, analysis of news media coverage, and participant observation should generate additional insights into medical tourism companies in Canada. For example, quantitative studies should provide insight into how many Canadians go abroad for care, where they seek treatment, and what types of procedures they select. Identifying Canada's medical tourism companies, providing their websites, describing their operations and business models, charting where they send patients abroad, summarizing the procedures they advertise, and explicating what services they offer in addition to organizing health care packages should facilitate further studies into Canada's medical tourism industry. It should also be of practical use to scholars interested in comparing medical tourism companies in Canada with similar businesses situated in other countries. For example, researchers in Australia, England, the United States and elsewhere might find it helpful to compare and contrast Canada's medical tourism industry with the emergence of medical tourism facilitators in their home countries. In addition, these findings should assist scholars interested in evaluating the quality of health-related information presented on the websites of medical tourism companies, and the use of YouTube, Twitter, Facebook, and various forms of social media by medical tourism companies situated in Canada. Scholars interested in using various research methods to study medical tourism companies in Canada should now have a better sense of where to focus their attention. To facilitate research by other scholars, websites or printouts of websites analyzed in this article have been electronically archived in a publicly accessible database.

6. LIMITATIONS

6.1 Findings limited to medical tourism companies based in Canada

This article is based upon content analysis of websites of medical tourism companies with home offices or affiliate offices in Canada. It does not provide insight into medical tourism companies located in other nations.

This study offers a set of categories or framework that can be used to study medical tourism companies in other nations. However, using content analysis to study websites of medical tourism companies based outside Canada might generate results very different from this study's findings.

6.2 Possibility of undiscovered websites

Internet searches, review of news reports, review of government records of federally incorporated companies in Canada, and examination of existing lists of medical tourism companies in Canada were all used in an attempt both to identify the total number of medical tourism companies established in Canada and distinguish medical tourism companies that are currently functioning from companies that are no longer involved in sending clients abroad for medical care. However, it is possible that some medical tourism companies were not identified in these searches. Medical tourism companies use company websites, Twitter, YouTube, Facebook, and other media to market international care, and internet searches are therefore a useful tool to use when trying to find medical tourism companies. Nonetheless, it is conceivable that some medical tourism companies based in Canada do not advertise using the internet or that repeated searches overlooked some businesses.

6.3 Possibility that non-responders might remain in business

Companies that did not respond to seven phone calls and/or emails over a three week period of repeated queries were listed as defunct. Consultants to the medical tourism industry recommend that calls to medical tourism companies should be promptly returned within twenty-four hours and preferably within two hours [52]. However, it is possible that some medical tourism companies continue to function even though they did not reply to multiple efforts to contact them. Establishing that businesses have ceased functioning in the absence of obvious signs such as companies losing their status as federally incorporated businesses in the end involved exercising judgement.

6.4 Investigative research methods might challenge claims made by medical tourism company websites

The study is limited insofar as it analyzes content found on medical tourism company websites and does not question the veracity of various claims. For example, one medical tourism company website states that clients can select clinics from among seven different countries. Someone using the tools of investigative journalism might discover that companies do not have business relations with all the facilities that they mention as international health care destinations. Such questions were bracketed from this study but they are worth pursuing in other studies. It is conceivable that some companies make exaggerated claims about destination facilities and advertised medical services and do not have experience coordinating particular health care packages and sending clients to advertised destinations. Content analysis, for the purpose of this article, accepted statements made by medical tourism companies. Such an approach, though it provides insight into how medical tourism companies in Canada advertise their services, would benefit from being complemented by investigative approaches that critically examine the credibility of claims made by medical tourism companies.

6.5 Studying Defunct Companies

This article focuses upon medical tourism companies that ceased operations. Performing a content analysis of websites of defunct companies could be seen as a limitation of the study, though the point of the exercise was to investigate and analyze not only medical tourism companies that are flourishing in a highly competitive marketplace but also to analyze websites of companies that have ceased sending clients to international health care destinations.

7. COMPARISON WITH PRIOR WORK

7.1 Focus on Canada

To date, most studies of medical tourism companies have focused upon websites of medical tourism facilitators based in the United States. This paper makes a meaningful contribution to scholarship by providing content analysis of medical tourism companies with home offices or affiliate branches in Canada. The local, regional, and national settings within which medical tourism companies are situated matters. It should not be assumed that medical tourism companies in Canada advertise the same health care packages or target the same prospective client base as businesses in Australia, the United States, and elsewhere.

7.2 Disclosure of Primary Sources

This article differs from previous work on medical tourism companies by identifying particular businesses, performing content analysis of their websites, and enabling readers to compare this analysis with the primary sources of medical tourism company websites. This step, by linking content analysis to identified primary sources, promotes transparency in research and allows other scholars to corroborate or challenge findings described in this study. Identifying particular companies and providing content analysis of their websites might help other researchers use their disciplinary methods and theories to study Canada's medical tourism industry.

7.3 Analyzing Company Websites

This article can be distinguished from most prior work on medical tourism due to its emphasis on content analysis of medical tourism company websites. Medical tourism companies conduct much of their business using the internet [53, 54, 55]. Reliance on the internet as a medium for communication, marketing, and interaction with clients means that much can be learned about medical tourism companies through careful analysis of their websites. Scholarly analysis of medical tourism companies should not be confined to content analysis of medical tourism company websites. Nonetheless, this mode of research makes an important contribution to the study of transnational medical travel.

7.4 Studying Inoperative Medical Tourism Companies

Finally, to date scholarship on medical tourism companies has examined companies currently engaged in the business of sending clients to international health care destinations. This article differs from previous work by identifying medical tourism companies that have ceased functioning. Identifying medical tourism companies that are inoperative raises questions about the nature and future direction of this industry. Many commentators assume that the medical tourism industry is in a period of unprecedented growth [56]. This article suggests that while the overall medical tourism industry might be expanding, the long-term success of individual medical tourism companies is not assured. The closure of twenty-five medical tourism companies in Canada might reflect a period of industry consolidation in which some companies flourish while others disappear or are subsumed by other companies. It is also possible that interest within Canada for travelling abroad to international medical facilities is not as large as is commonly assumed and these companies ceased functioning because many businesses are competing for a limited client base. Another possibility is that many of the companies erred when they based their operational model on sending clients to just one medical facility or country. It is conceivable that prospective clients want to have increased choice in where they travel for care and a business model that constrains options serves to deter clients that might otherwise consider going abroad for health care. And finally, since individuals considering going abroad for medical care can use the Internet to gain direct access to international hospitals and clinics, it is possible that medical tourism companies play a relatively minor role in the overall marketplace for medical travel. There might be less of a need for the service of medical travel facilitation than many entrepreneurs assume. Future research will explore in greater detail why the medical tourism companies identified in this study ceased functioning.

8. CONCLUSIONS

This study identifies twenty-five defunct Canadian medical tourism companies. Researchers will likely disagree concerning how to interpret some aspects of these findings. Individuals supportive of the Canada Health Act and publicly funded access to medically necessary health services and critical of the privatized, for-profit health care marketed by medical tourism companies might argue that the failure of twenty-five medical tourism companies is a sign of limited interest within Canada for medical travel. They might claim that news media coverage of the phenomenon of medical tourism far outpaces the actual significance of the topic. In contrast, proponents of for-profit, private health care might assert that in every competitive marketplace some companies prosper while other businesses disappear. Whatever lessons can be extracted from the closure of twenty-five medical tourism companies established in Canada, it seems reasonable to suggest that establishing a medical tourism company in Canada provides no assurance of commercial success.

This study accomplishes several goals. Medical tourism companies use the internet to market health services to prospective clients. Their websites provide valuable information about where medical tourism companies are located within Canada, what countries and facilities they select as destination sites, what health services they advertise, how they position themselves in a competitive national and global marketplace, and what travel-related services they promote in addition to marketing health care. Content analysis of medical tourism company websites permits development of a detailed account of the current state of Canada's medical tourism industry. This article reviews defunct medical tourism facilitators in Canada. Websites of operational medical travel companies based in Canada are examined elsewhere. Various research methods and theories can be used to build upon the analysis provided here.

Many researchers lament the lack of empirical research into the study of medical tourism. This article addresses these concerns and opens new avenues for the study of medical tourism companies, medical travelers, and international health care destinations. It also reveals the importance of studying medical tourism in Canada and elsewhere, and not making the error of thinking that the medical tourism industry is primarily a product of the large population of uninsured individuals in the United States. Many researchers and other interested parties will likely be surprised to learn just how many medical tourism companies have emerged in Canada, a country where residents have publicly funded access to a large basket of medically necessary health services. They might be equally surprised to learn how many of these businesses have come and gone and no longer market transnational medical travel to Canadians. Establishing just how many residents of Canada use the services promoted by medical tourism companies based in Canada, examining what happens to medical travelers when they go abroad for care, and learning more about their experiences once they return to Canada will have to await additional studies by researchers from a variety of disciplines. This study attempts to make a substantial contribution to the study of the Canadian medical tourism industry. It also provides sign posts that should be of practical use to health researchers and other individuals interested in exploring ethical, legal, social, cultural, economic, and public health implications of Canada's medical tourism industry.

AUTHOR'S CONTRIBUTIONS

LT is the sole author of this article. He designed the study, conducted content analysis of medical travel company websites, and wrote the manuscript.

ACKNOWLEDGEMENTS AND FUNDING

Study of Canada's medical tourism industry was facilitated by a Distinguished Visiting Fellowship (2006-2007) provided by the Comparative Program on Health and Society, Munk School of Global Affairs, University of Toronto. The Lupina Foundation provides funding for the Comparative Program on Health and Society. Analysis of Canadian medical tourism companies was also enabled by a William Dawson Scholar award (2006-2008) and additional sources of funding provided by McGill University. I wish to express appreciation for these sources of financial support. In addition, I wish to thank former colleagues at the University of Toronto as well as within the Biomedical Ethics Unit at McGill.

REFERENCES

1. Connell J: Medical tourism: Sea, sun, sand and...surgery. Tourism Management. 2006, 27: 1093-1100.

2. Garcia-Altes A: The development of health tourism services. Annals of Tourism Research. 2005, 32 (1): 262-266.

3. Goodrich JN, Goodrich GE: Health care tourism-an exploratory study. Tourism Management. 1987, 217-222.

4. Healy C: Surgical tourism and the globalization of healthcare. Ir J Med Sci. 2009, 178: 125-127.

5. Crooks V, Snyder J: Medical tourism: What Canadian family physicians need to know. Canadian Family Physician. 2011, 57: 527-529.

6. MacReady N: Developing countries court medical tourists. Lancet. 2007, 369: 1849-1850.

7. Ramirez de Arellano AB: Patients without borders: The emergence of medical tourism. IJHS. 2007, 37 (1): 193-198.

8. Unti JA: Medical and surgical tourism: The new world of health care globalization and what it means for the practicing surgeon. Bulletin of the American College of Surgeons. 2009, 94 (4): 18-25.

9. Behrmann J, Smith E: Top 7 issues in medical tourism: challenges, knowledge gaps, and future directions for research and policy development. Global Journal of Health Science. 2010, 2 (2): 80-90.

10. Lunt N, Carrera P: Medical tourism: assessing the evidence on treatment abroad. Maturitas. 2010, 66: 27-32.

11. Priest L: Sold! Triple bypass surgery in U.S. for $16,000. The Globe and Mail. 2011

12. Hopkins L, Labonte R, Runnels V, Packer C: Medical tourism today: what is the state of existing knowledge?. Journal of Public Health Policy. 2010, 31: 185-198.

13. Crooks VA, Kingsbury P, Snyder J, Johnston R: What is known about the patient's experience of medical tourism? A scoping review. BMC Health Services Research. 2010, 10: 266

14. Carrera P, Bridges JF: Globalization and healthcare: understanding health and medical tourism. Expert Rev. Pharmacoeconomics Outcomes Res. 2006, 6 (4): 447-454.

15. Kaspar C: A new lease on life for spa and health tourism. Annals of Tourism Research. 1990, 17 (2): 298-299.

16. Turner L: 'Medical tourism' and the global marketplace in health services: U.S. patients, international hospitals, and the search for affordable health care. IJHS. 2010, 40 (3): 443-467.

17. Milstein A, Smith M: America's new refugees-seeking affordable surgery offshore. NEJM. 2006, 355 (16): 1637-1640.

18. Schiano TD, Rhodes R: The dilemma and reality of transplant tourism: an ethical perspective for liver transplant programs. Liver Transplantation. 2010, 16: 113-117.

19. Kiatpongsan S, Sipp D: Monitoring and regulating offshore stem cell clinics. Science. 2009, 323: 1564-1565.

20. Smith E, Behrmann J, Martin C, Williams-Jones B: Reproductive tourism in Argentina: clinic accreditation and its implications for consumers, health professionals and policy makers. Developing World Bioethics. 10 (2): 59-69.

21. Huxtable R: The Suicide tourist trap: compromise across boundaries. Journal of Bioethical Inquiry. 2009, 6 (3): 327-336.

22. Turner L: 'Medical tourism' initiatives should exclude commercial organ transplantation. J R Soc Med. 2008, 101 (8): 391-394.

23. Burkett L: Medical tourism: concerns, benefits, and the American legal perspective. Journal of Legal Medicine. 2007, 28: 223-245.

24. Mattoo A, Rathindran R: How health insurance inhibits trade in health care. Health Affairs. 2006, 25 (2): 358-368.

25. Smith RD, Chanda R, Tangcharoensathien V: Trade in health-related services. Lancet. 2009, 373: 593-601.

26. Crozier GK, Baylis F: The ethical physician encounters international medical travel. J Med Ethics. 2010, 36: 297-301.

27. Turner L: 'First world health care at third world prices': Globalization, bioethics and medical tourism. BioSocieties. 2007, 2: 303-325.

28. Whittaker A, Manderson L, Cartwright E: Patients without borders: understanding medical travel. Med Anthropol. 2010, 29 (4): 336-343.

29. Johnston R, Crooks VA, Snyder J, Kingsbury P: What is known about the effects of medical tourism in destination and departure countries? A scoping review. International Journal for Equity in Health. 2010, 9: 24-10

30. Alleman BW, Luger T, Reisinger HS, Martin R, Horowitz MD, Cram P: Medical tourism services available to residents of the United States. J Gen Intern Med. 2011, 26 (5): 492-497.

31. Shah KY: Current status and emerging trends of international medical outsourcing in the United States--a qualitative study. 2007, University of South Carolina doctoral dissertation

32. Sobo EJ, Herlihy E, Bicker M: Selling medical travel to US patient-consumers: the cultural appeal of website marketing messages. Anthropology & Medicine. 2011, 18 (1): 119-136.

33. Johnston R, Crooks VA, Adams K, Snyder J, Kingsbury P: An industry perspective on Canadian patients' involvement in Medical Tourism: implications for public health. BMC Public Health. 2011, 11: 416-10.

34. Turner L: Medical tourism: Family medicine and international health-related travel. CFP. 2007, 53: 1639-1641.

35. Snyder J, Crooks V, Adams K, Kingsbury P, Johnston R: The patient's physician one-step removed': the evolving roles of medical tourism facilitators. Journal of Medical ethics. 2011, 37: 530-534.

36. Blackwell T: Cuba draws Canadian patients: Orthopedic, visual operations bring foreigners. National Post. 2007, A9-

37. McMurdy D: Vanity takes a vacation. National Post. 2007, FW.2.

38. Raman N: Medical travel agency lifts off. Montreal Gazette. 2006

39. McGinn D: Sun, sand and surgery. Globe and Mail. 2008

40. Ward D: Long wait times drive Canadians overseas for medical treatment. Edmonton Journal. 2007

41. Floyd M, Izenberg D, Kelly B, et al: Medical services directory. Maclean's. 2006

42. Find Medical Travel Services. [http://www.findinghealthcare.ca/medical_tourism.html]

43. Medical Tourism Facilitators. Tourism Review.com. 2009, [http://www.tourism-review.com/fm956/m6.pdf]

44. Pope C, Ziebland S, Mays N: Qualitative research in health care: Analysing qualitative data. BMJ. 2000, 320: 114-116.

45. Industry Canada. Corporations Canada, [https://www.ic.gc.ca/app/scr/cc/CorporationsCanada/fdrlCrpSrch.html;jsessionid=0000v9O3VPcCrmqiXhaZKPaOE2o:15dmaulj8?locale=en_CA]

46. Tri-Council Policy Statement 2. [http://www.pre.ethics.gc.ca/pdf/eng/tcps2/TCPS_2_FINAL_Web.pdf]

47. Turner L: Canadian Medicare and the Global Health Care Bazaar. Policy Options. 2007

48. Keckley PH, Underwood HR: Medical Tourism: Consumers in Search of Value. 2008, Deloitte Center for Health Solutions, [http://www.deloitte.com/assets/Dcom-UnitedStates/Local%20Assets/Documents/us_chs_MedicalTourismStudy(3).pdf]

49. Johnson TJ, Garman AN: Impact of medical travel on imports and exports of medical services. Health Policy. 2010, 98 (2-3): 171-177.

50. Puzic S: Cancer patient bilked in alleged fraud. Windsor Star. 2010

51. Puzic S: Windsor police lay fraud charges against EcuMedical couple. Windsor Star. 2010

52. Passmore R: Converting international patient inquiries @10%-Part 2. Medical Tourism Magazine. 2009

53. Cormany D, Baloglu S: Medical travel facilitator websites: An exploratory study of web page contents and services offered to the prospective medical tourist. Tourism Management. 2011, 32: 709-716.

54. Lunt N, Hardey M, Mannion R: *Nip, tuck and click*: Medical tourism and the emergence of web-based health information. The Open Medical Informatics Journal. 2010, 4: 1-11.

55. Mason A, Wright KB: Framing medical tourism: An examination of appeal, risk, convalescence, accreditation, and interactivity in medical tourism web sites. Journal of Health Communication. 2011, 16 (2): 163-177.

56. York D: Medical tourism: The trend toward outsourcing medical procedures to foreign countries. Journal of Continuing Education in the Health Professions. 2008, 28 (2): 99-102.

Chapter 7

Beyond "medical tourism": Canadian companies marketing medical travel

Leigh Turner

Center for Bioethics, School of Public Health, College of Pharmacy, University of Minnesota, Minneapolis, MN, USA

ABSTRACT

Background

Despite having access to medically necessary care available through publicly funded provincial health care systems, some Canadians travel for treatment provided at international medical facilities as well as for-profit clinics found in several Canadian provinces. Canadians travel abroad for orthopaedic surgery, bariatric surgery, ophthalmologic surgery, stem cell injections, "Liberation therapy" for multiple sclerosis, and additional interventions. Both responding to public interest in medical travel and playing an important part in promoting the notion of a global marketplace for health services, many Canadian companies market medical travel.

Methods

Research began with the goal of locating all medical tourism companies based in Canada. Various strategies were used to find such businesses.

During the search process it became apparent that many Canadian business promoting medical travel are not medical tourism companies. To the contrary, numerous types of businesses promote medical travel. Once businesses promoting medical travel were identified, content analysis was used to extract information from company websites. Company websites were analyzed to establish: 1) where in Canada these businesses are located; 2) the destination countries and health care facilities that they market; 3) the medical procedures they promote; 4) core marketing messages; and 5) whether businesses market air travel, hotel accommodations, and holiday tours in addition to medical procedures.

Results

Searches conducted from 2006 to 2011 resulted in identification of thirty-five Canadian businesses currently marketing various kinds of medical travel. The research project began with what seemed to be the straightforward goal of establishing how many medical tourism companies are based in Canada. Refinement of categories resulted in the identification of eighteen businesses fitting the category of what most researchers would identify as medical tourism companies. Seven other businesses market regional, cross-border health services available in the United States and intranational travel to clinics in Canada. In contrast to medical tourism companies, they do not market holiday tours in addition to medical care. Two companies occupy a narrow market niche and promote testing for CCSVI and "Liberation therapy" for multiple sclerosis. Three additional companies offer bariatric surgery and cosmetic surgery at facilities in Mexico. Four businesses offer health insurance products intended to cover the cost of obtaining privately financed health care in the U.S. These businesses also help their clients arrange treatment beyond Canada's borders. Finally, one medical travel company based in Canada markets health services primarily to U.S. citizens.

Conclusions

This article uses content analysis of websites of Canadian companies marketing medical travel to provide insight into Canada's medical travel industry. The article reveals a complex marketplace with different types of companies taking distinct approaches to marketing medical travel.

KEYWORDS

Medical tourism; Canada; Transnational healthcare; Globalization; Websites

1. BACKGROUND

In 2006 I began studying what many journalists and health researchers label "medical tourism". A Canadian citizen, I started my research with particular interest in studying medical tourism companies based in Canada. I wanted to better understand why some Canadians seek medical procedures outside Canada. In addition, from the start of my research I examined ethical issues related to medical travel and the emergence of a global marketplace in health services [1, 2, 3, 4]. As of 2011 I have identified sixty-three businesses that have head offices or affiliate offices in Canada and market health care at medical facilities based outside Canada. Of this total, twenty-eight companies are no longer operational. The remaining thirty-five companies continue marketing health care delivered at international facilities.

To provide insight into Canada's medical travel industry, I use content analysis of company websites to study the many different businesses that are based in Canada and market access to health care provided at international destinations as well as in for-profit clinics in Canada. Before conducting content analysis of medical travel company websites I first had to identify all companies that are based in Canada and market medical travel to international health care facilities. From 2006 to 2011 I developed a database of Canadian businesses promoting medical travel. I began with the assumption that I would identify and describe Canadian medical tourism companies. However, as I studied various businesses, I noticed important differences in the types of companies marketing transnational and intranational health care.

Some Canadian companies market health care in such countries as Costa Rica, India, and Thailand; they also offer holiday tours and stays at resorts. These companies typically promise access to treatment at distant health care facilities, offer to arrange air travel and hotel accommodations, list different medical procedures, transfer medical records, and promote "all-inclusive" medical tourism packages ensuring access to affordable, timely, and high-quality care. These businesses fit the model of what most researchers characterize as medical tourism companies.

Other businesses in Canada promote medical travel but do not send their clients to distant international destinations. Rather, these companies promote regional, cross-border travel to hospitals and clinics in the United States. They also market intranational travel to for-profit clinics within Canada. There is a long history of cross-border medical travel between Canada and the United States; some companies attempt to facilitate such travel [5, 6, 7, 8]. In contrast to many medical tourism companies, these companies are more focused on facilitating prompt access to medical procedures than combining health care with holidays.

In addition to identifying medical tourism companies and cross-border medical travel facilitators, I found numerous businesses occupying specific niches within the medical travel industry. For example, I found companies that specialize in coordinating access to testing for Chronic Cerebrospinal Venous Insufficiency (CCSVI) and "Liberation Therapy" for multiple sclerosis [9]. Other businesses market access to bariatric surgery and cosmetic surgery performed at international destinations. Also, I found Canadian businesses that market insurance products intended to promote access to health care facilities in the U.S [10, 11]. These businesses offer critical illness insurance plans and other insurance products and also help their clients coordinate care outside Canada. Finally, I found one company that is based in Canada and yet markets medical procedures predominantly to U.S. citizens. I began my research by looking for Canadian medical tourism companies and in time discovered a marketplace considerably more complicated than I anticipated. I have tried to describe these different businesses and sort them into distinct categories.

To transform into a manageable endeavor analysis of such a large number of companies, I gathered data on basic features of these businesses. First, I identified companies and established where they are located within Canada. Second, I recorded destination nations and, where noted, particular health care facilities promoted by medical travel facilitators. Third, I identified the medical procedures and medical specialties businesses market. Fourth, I extracted core marketing messages of medical travel companies. Fifth, I analyzed company websites to determine whether businesses offer to book travel, make hotel reservations, and coordinate tours and side trips in destination nations.

2. METHODS

2.1 Development of a database of medical travel companies located in Canada

Before conducting content analysis of company websites I first developed a database of businesses marketing medical travel. Numerous methods were used to build the database. From 2006 to July 2011, I conducted repeated Internet searches to identify medical travel companies with head offices or affiliate offices in Canada. In total I identified thirty-five companies currently promoting medical travel beyond Canada's borders and 28 companies that have exited the marketplace for promoting transnational health care.

When conducting Internet searches I used such phrases as "medical tourism Canada", "medical tourism company Canada", medical tourism agency Canada" medical tourism facilitator Canada", "medical tourism broker Canada", "medical tourist Canada", and "cross-border healthcare Canada". I also used these terms when searching for newspaper articles describing Canadian medical travel companies. Searches for newspaper articles were conducted using Google News Canada and ProQuest Newsstand. Newspaper reports of medical tourism companies assisted with tracking the development of Canadian businesses marketing medical travel. Use of Google Alerts complemented searches for relevant newspaper articles. Google Alerts were created for such phrases as "medical tourism Canada", "stem cell tourism Canada", "transplant tourism Canada", and "global health care Canada". Industry Canada operates a website that can be searched to identify federally incorporated Canadian businesses [12]. Searches of this database revealed the identities of several Canadian medical tourism companies. In addition, I was able to identify additional businesses and assess the comprehensiveness of the database by discovering brief lists of Canadian medical travel companies [13, 14, 15]. Continuing my search for Canadian companies involved in promoting medical travel, in November 2009 I traveled to Toronto, Ontario and attended an exhibition and conference on medical tourism [16]. In short, multiple search strategies were used in an effort to find Canadian businesses engaged in marketing medical travel.

Phone calls and emails were used to establish whether companies continue to function. Companies were deemed to have exited the marketplace if they had expired websites, non-functioning email accounts and disconnected phone service, or seven phone calls and/or emails failed to elicit a response. Companies were deemed operational if respondents reported that the companies remain in business.

2.2 Content analysis of websites of medical travel companies

Once medical travel companies were identified, content analysis was used to study the websites of these businesses [17]. Content analysis was performed by analyzing printed versions of company websites. There were five main elements to content analysis of company websites. First, content analysis was used to identify the province and city or town that company websites indicate as their primary address within Canada. Where affiliate offices within or outside Canada were noted, I recorded these locations as secondary sites. Acquiring this information makes it possible to establish where across Canada medical travel companies are based. Medical travel companies have, in effect, two "locations". In some respects they are internet-based companies that occupy virtual space [18, 19]. Acknowledging the powerful role of the Internet as a tool for marketing transnational health care, the geographic location of medical travel companies matters. In Canada, medical travel companies are bound by the laws and regulations of both the provinces in which they are based and by federal law. For the purpose of enforcing laws and regulations, the location of companies is consequential [20].

Second, I used content analysis to document countries and, where listed, the specific health care facilities that medical travel companies identify as potential destination sites for their clients. Some company websites list a single destination for prospective customers. Other companies provide lengthy lists of possible destinations for health care. Websites were analyzed for the purpose of trying to understand where Canadian medical travel companies offer to send their clients.

Third, I used content analysis to document medical procedures and/or medical specialties marketed by medical travel companies. Some company websites list such specialties as cardiology and orthopaedics. Other company websites provide lengthy lists of particular medical procedures. Yet other websites list both clinical specialties and particular medical interventions. This topic was addressed to establish the types of health services medical travel companies market.

Fourth, I reviewed company websites for core marketing statements. These statements were found on home pages of medical travel companies, in mission statements, and on pages labeled with such tags as "About Us" and "What We Do". Marketing messages were then summarized to indicate whether companies emphasize such messages as affordable access to care, access to timely health services, access to high-quality care, and access to care in exotic destinations.

Fifth, content analysis was used to establish whether medical travel companies market air travel, book hotel reservations, and arrange side trips to local tourist sites and other holiday excursions in addition to marketing health services. This subject was addressed to explore how many medical travel companies promote tourism and travel-related services in addition to health care.

Detailed information was recorded for each category of analysis. Information extracted through content analysis was fact-checked and feedback was solicited from two senior colleagues. Readers can assess the quality of content analysis by comparing it to the primary data of medical travel company websites. All company names are disclosed, website addresses are provided, and company websites are electronically archived.

Following the initial round of content analysis, phone calls, emails, Internet searches and searches of an Industry Canada database were used to distinguish operational medical travel companies from businesses that have exited the marketplace for transnational health care. Companies with expired websites, non-functional and non-responding email addresses, and discontinued phone numbers were deemed to have exited the marketplace. Businesses that have exited the marketplace are the subject of a companion article. This article describes and analyzes operational Canadian medical travel companies.

3. RESULTS

3.1 Types of medical travel companies

I began studying Canada's medical travel industry with what I assumed was a straightforward plan to identify all medical tourism companies based in Canada. During the process of trying to establish how many businesses constitute Canada's medical tourism industry, it became apparent that companies operate according to somewhat different business models. Of the thirty-five companies involved in coordinating medical travel, eighteen Canadian companies fall within the category of what most health researchers regard as medical tourism companies [21, 22, 23, 24]. These companies promote health services provided in such international destinations as Barbados, Costa Rica, India, Mexico, and Thailand. Seven additional companies share much in common with the eighteen medical tourism companies promoting medical travel to distant locations. However, instead of marketing health services in such distant countries as India or Mexico they restrict themselves to promoting regional, cross-border travel to the United States as well as private clinics within Canada. These companies, could, with some justification, be included in the category of medical tourism companies. However, it is helpful to note that some medical travel companies promote regional, cross-border travel, as well as intranational travel within Canada, instead of marketing travel to distant international health care facilities. These companies emphasize timely access to care rather than promoting holiday excursions. Two businesses promote medical travel but restrict themselves to marketing testing for CCSVI and "Liberation therapy" for multiple sclerosis. Three additional businesses advertise bariatric surgery and cosmetic surgery performed at sites in Mexico. Four businesses market private health insurance products that permit Canadian citizens to obtain access to health care in the U.S. and Canada. These companies also help arrange care at medical facilities in the U.S. and private Canadian clinics. Finally, one medical travel company markets health services primarily to uninsured and underinsured U.S. citizens. Medical travel companies can be sorted into different categories even though they all promote transnational health care. They occupy distinct market niches, promote different kinds of medical procedures, and offer different kinds of services to prospective clients. This article identifies different types of companies marketing medical travel and does not attempt to use "medical tourism" as a catch-all term for describing every business involved in promoting transnational and intranational medical travel [25].

3.2 Locations of medical travel companies

Of the eighteen medical tourism companies, five are based in British Columbia, four are located in Alberta, one is based in Saskatchewan, one is in Manitoba, four are in Ontario, and three are based in Quebec.

Seven companies market cross-border, regional medical travel to the U.S. as well as intranational medical travel within Canada, Two of these businesses are situated in British Columbia, four are based in Ontario, and one is located in Quebec.

The two companies marketing diagnostic testing for CCSVI and Liberation Therapy for multiple sclerosis are based in Manitoba and Ontario.

Three companies advertise out-of-country bariatric surgery and cosmetic surgery. One business is located in Alberta. The remaining two companies are based in Saskatchewan.

Four businesses market private health insurance products that enable access to U.S. health care facilities. Two of these companies are located in Alberta, one is based in Ontario, and one is located in Quebec. In addition to offering insurance products these companies help their clients arrange care in the U.S.

The lone Canadian medical travel company marketing primarily to U.S. citizens is located in British Columbia.

Some medical travel companies based in Canada identify affiliate offices or company representatives located outside Canada. Where company websites note such information I have recorded where company representatives situated outside Canada are based. Additional file 1 lists company names, identifies where in Canada these businesses are located, and indicates affiliates and company representatives in those instances where companies had offices or agents situated outside Canada.

3.3 Destination countries and health care facilities

Of the eighteen medical tourism companies promoting global medical travel, seven marketed one country as a health care destination, two businesses marketed three health care destinations, three companies marketed five countries as destinations, one company marketed six destination nations, two companies marketed eight destination nations, one company marketed nine destination nations, one business marketed twelve destination nations, and one business listed thirty potential destination nations.

In total, the eighteen medical tourism companies promoting global medical travel list thirty-eight different destination nations. Eleven companies list India as a potential medical destination, nine companies list Coast Rica, seven companies list Thailand, seven list Mexico, six companies list the U.S., five companies list Singapore, five companies list Canada, three companies list South Africa, three companies list Brazil, three companies list Malaysia, three companies list Turkey, three companies list Barbados, two companies list Italy, two list Cuba, two list the Dominican Republic, two list Poland, two list Argentina, two list Panama, two list Colombia, two list Israel, and two list the United Arab Emirates (with one company indicating Dubai as a potential destination and one company identifying Abu Dhabi as a possible destination site). El Salvador, United Kingdom, Ukraine, Switzerland, Spain, South Korea, Ecuador, France, Austria, Bahamas, Belgium, Bulgaria, Germany, Hungary, Jordan, Lebanon, and Peru are all listed by a single medical tourism company. The list of possible destination nations, and the number of companies that mention them, reveals the variety of destinations marketed by medical tourism companies based in Canada as well as the most common destination nations promoted by Canadian medical tourism companies.

Of the seven companies marketing cross-border, regional travel to the U.S. and intranational travel to clinics within Canada, six companies indicate that tests and procedures can be obtained in both countries and one business restricts itself to sending clients to U.S. medical facilities. The lone company marketing medical travel primarily for diagnostic imaging is among the businesses offering access to health services in both the U.S. and Canada.

Of the two companies marketing medical travel for “CCSVI testing” and “Liberation therapy”, one business marketed diagnostic imaging in the U.S. and procedures in India. The other company markets both tests and procedures in India.

Three businesses market medical travel for bariatric surgery and cosmetic surgery. All of these companies advertise surgery performed in Mexico.

Four businesses market private insurance plans that enable access to health care in the United States. Three of the businesses note the possibility of obtaining some types of tests and treatments at medical facilities in Canada. One insurance program is restricted to enabling access to tests and procedures at three Mayo Clinic sites based in the U.S.

Finally, one medical travel company based in Canada targets U.S. customers rather than Canadians interested in travelling for medical care. This business markets access to medical procedures available in eight U.S. states as well as facilities in Canada.

It is important to make a distinction between how medical travel is marketed on company websites and actual practices of Canadians using services of medical travel businesses when going abroad for health care. When medical travel companies list various countries as potential destinations for their clients, it should not be assumed that Canadians have contracted with these companies and travelled to listed destinations. It is conceivable that clients of these businesses have never selected some listed countries. It is also possible that some nations are listed even though companies have no established relationships with health care providers in these countries. Acknowledging the possibility of a gap between how medical travel is marketed and behaviour of medical travelers, analyzing websites of Canadian medical travel companies provides insights into how these businesses attempt to market health services to prospective clients. Additional file 1 identifies locations of destination facilities marketed on company websites.

3.4 Types of advertised medical procedures

Most of the 18 medical tourism companies in Canada offer comprehensive baskets of health services. However, some companies take niche positions by advertising restricted range of medical interventions. For heuristic purposes, medical travel companies can be classified as "generalist", "specialist", and "intermediate" firms. The former category includes businesses offering, for example, orthopaedic procedures, infertility treatments, cosmetic surgery, cardiac care, ophthalmology procedures, alternative medicine, and dental surgery. A company limited to marketing cosmetic surgery procedures can serve as an example of a specialist medical tourism company. Of the 18 medical tourism companies promoting health care at global destinations, 14 businesses operate as generalist medical travel firms. Two companies are specialist firms. One of them specializes in cosmetic surgery. The second firm predominantly markets procedures related to administration of human umbilical cord blood cells. Two businesses fall between these poles and can be classified as having "intermediate" marketing models. Of these latter businesses, one company emphasizes in its marketing claims access to cosmetic surgery but its list of procedures includes cosmetic surgery, dental surgery, bariatric surgery, gender-reassignment surgery, and non-surgical interventions. Another firm claims that it primarily markets alternative health care but also advertises orthopaedic surgery and reconstructive surgery.

Of the seven companies marketing cross-border, regional medical travel to health care facilities in the United States as well as private clinics in Canada, four companies can be classified as generalist firms. Of the remaining three businesses, one company promotes access to many different kinds of care but places particular emphasis upon diagnostics and arranging second opinions, one company offers comprehensive services but emphasizes preventive medicine and access to diagnostic tests, and one business specializes in marketing diagnostic imaging.

Of the remaining medical travel companies, two businesses market testing for CCSVI and “Liberation Therapy” for MS, and three businesses predominantly market bariatric surgery and cosmetic surgery. Both of the latter companies emphasize that they focus upon providing weight-loss treatments. These companies can all be classified as “specialist” medical travel firms.

Four companies market private health insurance plans enabling access to out-of-province and out-of country health care. These businesses offer insurance products that cover range of medical interventions. However, since their main product related to medical travel is related to selling critical illness insurance that can be used to cover cost of care at private medical facilities, they can be classified as specialist firms.

Finally, one medical travel company that is based in Canada markets health services mainly to clients from outside Canada. This company operates with a generalist model and offers such health services as cardiac surgery, general surgery, orthopedic surgery, neurosurgery, High Intensity Focused Ultrasound for prostate cancer, and additional medical interventions.

In summary, of thirty-five companies marketing medical travel, nineteen businesses operate using a generalist model based upon advertisement of many different kinds of medical interventions or insurance products and customer service intended to promote access to health care. Additional file1 lists health services and medical specialties promoted by Canadian medical travel companies.

3.5 Core marketing messages of medical travel companies

Most medical tourism companies promoting travel to global destinations use core marketing messages that emphasize access to affordable, timely, and high quality care. Of eighteen medical tourism companies marketing travel to global destinations, sixteen emphasize affordability of health care at international medical facilities. Access to timely health care is marketed by fifteen companies. All eighteen businesses market access to high-quality health care. Three businesses note that medical interventions can be obtained in exotic settings that appeal to tourists.

Of the seven businesses marketing regional, cross-border medical travel to the U.S. as well as intranational medical travel within Canada, six companies emphasize affordability of care. Six businesses also market timely access to treatment. Four companies emphasize the high-quality of care available at the destinations they promote.

Two companies market testing for CCSVI and Liberation Therapy. Both companies emphasize affordability and high-quality of marketed procedures; one business promotes timely access to treatment.

All three companies marketing bariatric surgery at facilities outside Canada emphasize comprehensive access to weight loss interventions. Their core marketing messages emphasize access to bariatric surgery and identify the variety of weight loss programs they offer. Their lists of specific procedures include cosmetic surgery.

Four businesses market insurance products that enable access to health care outside Canada or at private clinics within Canada. Three companies emphasize affordable access to care. All four companies promote timely access to health care and access to treatment at high-quality healthcare facilities.

Finally, the one medical travel company targeting non-residents of Canada promotes access to affordable, timely, and high-quality medical interventions.

There are variations in the types of themes companies use when marketing medical travel. However, messages concerning affordability of care, timely access to treatment, and access to high-quality medical interventions are common. Some companies use all three selling points to market health services at international destinations. Other businesses emphasize one or two key features of the health services they promote. Additional file 2 summarizes medical travel companies' core marketing messages.

3.6 "Tourism" component of medical travel

Medical travel companies typically emphasize affordability of care, timely access to care, and quality of care rather than tourism-related activities in their core marketing messages. Some companies note that they offer access to care in "exotic settings". However, incorporating rhetoric concerning holiday-going and tourism activities into core marketing messages is atypical. Acknowledging that medical travel companies place greater emphasis on advertising medical interventions than promoting tourism, many businesses nonetheless market services associated with travel and tourism. Of the eighteen companies marketing medical tourism at global health care destinations, thirteen offer to coordinate travel arrangements, seventeen advertise the service of booking hotel reservations or otherwise arranging accommodations for clients, and fourteen offer to organize tours to local attractions located near where medical procedures are provided.

Of the seven companies marketing cross-border regional travel to the U.S. and intranational travel within Canada, four offer the service of booking travel and three offer to make hotel reservations. The possibility of booking tours to local holiday destinations is not addressed by these company websites.

Two companies market testing for CCSVI and Liberation Therapy. Of these businesses, one offers to book travel, both offer to make hotel reservations, and one advertises tours.

Three companies market weight-loss surgery and cosmetic surgery at international destinations. All three businesses offer to book travel; two companies offer to make hotel reservations. None of the companies mentions organizing tours and side trips before or after medical care is provided to clients.

Of the four companies marketing insurance products enabling access to care in the United States as well as private clinics in Canada just one business clearly indicates that it books travel. None of the companies state whether they book hotel reservations and organize tours. These companies emphasize offering insurance products and ensuring that their clients can obtain timely access to care in the United States or Canada. They help arrange care at U.S. facilities but they do not organize logistics to the same extent as medical tourism companies and regional, cross-border medical travel companies. In addition, they do not promote tours and holiday excursions.

The one Canadian medical tourism company marketing care to U.S. citizens does not indicate whether it books travel, make hotel reservations, or organize tours and other holiday excursions. It facilitates medical procedures and does not advertise additional services.

In summary, there is considerable variation in the extent to which medical travel companies market coordination of air travel, hotel reservations, and holiday excursions in addition to promoting medical procedures. Some companies embrace the concept of medical tourism and combine marketing of medical procedures with holiday excursions in "exotic" locations. Other businesses focus exclusively on arranging medical care at international destinations and make no attempt to promote and organize travel, accommodations, and tours.

Additional file 2 identifies whether medical travel companies advertise travel arrangements, offer to organize hotel accommodations, and market tours and side trips in addition to marketing health services.

4. DISCUSSION

This article provides a content analysis of company websites of thirty-five Canadian businesses that market medical travel. Content analysis of company websites is a valuable tool for exploring what services companies promote and how they attempt to position themselves in the marketplace [26,27]. Drawing upon content analysis of websites of medical travel companies, it is reasonable to conclude that Canada now has many businesses marketing access to medical procedures that are paid for out-of-pocket, rather than covered by provincial health insurance plans, and provided at international medical facilities. At present, Canada has a limited private health care sector [28, 29]. Provincial health care systems provide universal access to "medically necessary" health services; legislation restricts the types of medical procedures for which private clinics are allowed to charge patients [30, 31]. Canada's effort to provide publicly-funded, universal access to health care is based on an egalitarian philosophy in which all citizens have rights to medical care and personal wealth cannot be used to jump to the front of queues and obtain treatment [32]. Legislation limiting expansion of a for-profit, private health care sector in Canada appears to have resulted in the emergence of companies that serve as bridges or intermediaries to hospitals and clinics located outside Canada. Some of these businesses also market intranational travel to private, for-profit medical clinics located in British Columbia, Alberta, Quebec, and elsewhere. Medical travel companies promote access to an out-of-country private health care sector available to Canadian citizens with sufficient financial resources. Various insurance products provide Canadians with sufficient financial resources tools for gaining access to health care facilities in the U.S.

Presumably wanting to appeal to the largest possible number of prospective clients, most medical travel companies offer many different kinds of medical procedures. Though Canadian provincial health insurance systems provide universal access to care for medically necessary services, Canadians must sometimes wait long periods for elective medical procedures [33, 34, 35]. Some medical travel companies respond to treatment delays in Canada by marketing swift access to private health care at international facilities. Orthopedic procedures, ophthalmologic procedures, bariatric surgery, and other interventions that sometimes require lengthy waits for care in Canada are readily available within the private health care sectors of other countries. Many Canadian medical tourism companies sending their clients to such countries as India and Thailand, as well as medical travel companies promoting regional, cross-border travel to the United States, in their core marketing messages emphasize timely access to care.

Assurances about providing timely access to care can be found on the websites of numerous medical travel companies based in Canada. Claims about offering access to high-quality care are also widespread. Clients of medical travel companies presumably want not just fast access to treatment but also access to professional medical care. Company websites offer messages intended to reassure clients about treatment available at international hospitals and clinics.

There are some circumstances where provincial health insurance plans in Canada permit reimbursement of expenses incurred as a result of receiving out-of-country medical care. In Canadian provinces, coverage of out-of-country care for elective procedures typically requires, at minimum, recommendation by a Canadian physician and pre-approval by provincial health ministries [36]. In most instances individuals travelling abroad for elective medical procedures must pay out-of-pocket for treatment [37]. Perhaps for this reason, many medical travel companies in Canada emphasize affordability of the medical procedures they market.

Some companies specialize in offering medical interventions that have not undergone clinical trials or regulatory review and are not approved for patient care in Canada. For example, medical travel companies marketing access to stem cell injections and "Liberation Therapy" help their clients gain access to procedures that patients cannot obtain in Canada. From one perspective, offering such services promotes "patient choice" by giving clients access to procedures they wish to undergo. However, expansion of choice likely also exposes some Canadians to considerable risk by enabling them to gain access to medical interventions with unknown safety and efficacy profiles. To date, at least two Canadians with multiple sclerosis are reported to have died while undergoing "Liberation Therapy" at medical facilities located outside Canada [38, 39, 40].

5. Limitations

Multiple search methods were used to identify Canadian medical travel companies. However, Canada's marketplace for medical travel is quite turbulent. New companies emerge, old businesses disappear, companies that have marketed medical travel for years periodically repackage themselves, drop old marketing initiatives, and fashion new identities. Due to limitations in search strategies and transformations in the marketplace for medical travel, it is possible that this article has not examined websites of all medical travel companies based in Canada. In addition, given fluidity in the marketplace, some companies that were operational when they were contacted might subsequently have gone out of business.

Internet searches and searches of databases containing newspaper articles were conducted in English. Though I was able to identify medical travel companies based in Quebec, it is conceivable that some companies marketing in French or other languages were not identified when using different search strategies.

Content analysis is useful when attempting to describe what companies claim to do. It is not a useful tool for attempting to establish whether there is a gap between promotional claims and actual company practices. Companies can engage in misrepresentation, use misleading advertising, and even defraud their customers by marketing health services and then failing to deliver what they are contractually obligated to provide [41, 42]. Interviews, participant observation, surveys, and tools drawn from investigative journalism can all provide insight into possible gaps between how companies use the internet to market themselves and how they function in practice. Future research will help address the extent to which marketing claims resemble or differ from business practices.

This study has an obvious geographic limitation. I examine medical travel companies based in Canada. Though methods used in this paper can be applied to the study of medical travel companies in other countries, I do not address medical travel companies based outside Canada. Indeed, the paper does not address all of the medical travel companies that Canadians might consult when deciding whether to seek care outside Canada [43]. Canadians considering going abroad for treatment are free to contact destination hospitals or select health care packages provided by medical travel companies based outside Canada. Studying medical travel companies within Canada sets reasonable limits for research, and I suspect many Canadians consider using Canadian medical travel businesses if they decide to use a facilitator to arrange transnational health care. However, it is important to note that companies based in the United States and elsewhere also market health services to Canadian citizens.

6. COMPARISON WITH PRIOR WORK

6.1 Focus on Canada

U.S. medical tourism companies are the subject of numerous studies [44, 45, 46]. Medical tourism companies based in Canada have attracted some attention from researchers but there is limited analysis of company websites and little exploration of the many medical travel companies that do not neatly fit into the category of medical tourism businesses [47, 48, 49]. This article examines the medical travel industry of a country with a publicly funded system of universal access to health care. There are presumably some important differences between why Canadians go abroad for health care and why residents of the United States and other countries travel for medical interventions. Scholarly research examining medical travel will benefit from analysis of many different social, economic, and cultural contexts.

6.2 Investigating "medical tourism companies" as a subset of medical travel businesses

To date, researchers appear particularly interested in studying companies that promote medical travel to India, Thailand, and other distant, "global" health care destinations. This article makes distinctions among different types of companies in Canada's medical travel industry. Some companies market health services available in countries located far from Canada. Other companies promote regional, cross-border travel to the U.S. as well as intranational travel within Canada. Yet other businesses promote specific interventions such as bariatric surgery and "Liberation therapy". Some companies market insurance products and help Canadian clients obtain health care in the U.S and private facilities in Canada. Studies that focus strictly on medical tourism companies sending clients to distant nations risk missing or neglecting many of the businesses identified in this study. Businesses promoting medical travel are not reducible to medical tourism companies marketing health services and holidays available in faraway international locations. There is something to be gained from identifying differences within the medical travel industry and expanding the analytic lens to include both medical tourism companies and related businesses that promote medical travel but do not necessarily fit the standard model of medical tourism companies. Businesses promoting regional, cross-border care, weight loss, "Liberation therapy" and private insurance coupled with assistance arranging health care in the United States all market access to various types of medical travel.

6.3 Canada's medical travel industry as a bounded field

Several studies analyze particular facets of medical tourism company websites but do not use search strategies designed to identify companies based in particular countries. This study attempts to identify and analyze websites of all identified operational medical travel companies situated in

Canada. Such an approach makes it possible to provide an overview of Canada's medical travel industry.

6.4 Disclosure of primary sources

This article identifies Canadian companies that market transnational health care, lists their websites, and electronically archives all company websites. Additional file 3 provides company names, notes their websites, and contains links to electronically archived copies of company websites. These steps mean that readers have access to all primary data used as the basis for this study. Access to primary data means that it is possible for readers to evaluate claims made in this paper, conduct novel analyses driven by research questions not addressed in this study, compare Canada's medical travel industry with medical travel companies in other countries, and track over time changes in Canada's medical travel industry. Identifying specific businesses and providing content analysis of company websites should facilitate further research into Canada's medical travel industry and enable comparative, cross-national studies.

7. CONCLUSIONS

Canada has a health care system providing what is often described as offering universal access to medically necessary care. Most citizens belong to publicly funded health insurance plans run by provinces. These plans are guided by the egalitarian philosophy that all citizens, regardless of social status and personal wealth, should have publicly funded access to "medically necessary" health services. In short, need for care rather than capacity to afford treatment is supposed to determine access to health care. Given that residents of Canada have universal access to medically necessary health services, individuals unfamiliar with health care in Canada might assume that with presumably limited demand few medical travel companies are likely to emerge in Canada. To the contrary, thirty-five businesses market medical travel.

Eighteen businesses market health care at such international destinations as Barbados, Costa Rica, India, and Thailand. Most of these companies offer wide range of health services. Many of these companies market air travel, hotel accommodations, and holiday excursions that can take place before or after receiving medical care. These companies take medical care and tourism and fuse them together to create a novel type of business enterprise.

Seven businesses market regional, cross-border health care. These companies advertise medical procedures that can be obtained in the United States and, in some instances, at private medical clinics in Canada. Though Canada has a private health care sector, it is limited in scope. In contrast, the U.S. has a large private health care sector in which it is possible to obtain ready access to care by paying out-of-pocket for treatment. Companies promoting regional, cross-border medical travel as well as intranational travel within Canada take advantage of the large private health care sector in the U.S. and the small but growing private health care sector in Canada [50, 51, 52]. In many respects they resemble the eighteen businesses identified as "standard" medical tourism companies. However, these businesses differ in terms of where they propose sending their clients. They also differ in the extent to which they promote tourism-and-travel-related activities in addition to marketing health services.

"Liberation therapy" and CCSVI as a diagnostic category for understanding multiple sclerosis receive extensive news media coverage in Canada. In response to public interest in Liberation Therapy, and perhaps also stoking interest in the procedure, two medical travel companies in Canada dedicate themselves exclusively to promoting access to testing for CCSVI and "Liberation Therapy". The success of their business model is likely connected to whether "Liberation Therapy" continues to be seen by some Canadians with multiple sclerosis as potentially effective medical intervention [53, 54].

Three companies market bariatric surgery performed at facilities based outside Canada. These companies are businesses that promote weight loss strategies; they are not exclusively dedicated to promoting medical travel for bariatric surgery. However, since these businesses send their clients to Mexico, I have categorized them among the different types of companies promoting medical travel. In Canada, patients must often face long waits for bariatric surgery. These companies have presumably responded to these delays by promoting prompt access to bariatric surgery at medical facilities beyond Canada's borders.

Four businesses market critical illness insurance products that help Canadians have the financial resources they will need to obtain private health care within the United States or at private clinics within Canada.

Across Canadian provinces there are numerous restrictions on private health insurance for medically necessary medical interventions [55]. These companies skirt these restrictions by providing insurance for health services delivered outside Canadian provincial health insurance plans. In addition to providing financial support for health care provided in the U.S. or Canada's private health sector, these businesses help their clients arrange care at U.S. medical facilities. However, it is important to appreciate that in numerous respects these businesses differ from medical tourism companies. They do not, for example, market holiday excursions before or after treatment. Rather, they offer private health insurance products that can be used to cover expenses associated with obtaining out-of-pocket medical care.

Finally, one medical travel company based in Canada markets health services primarily to U.S. citizens. The U.S. has a large population of uninsured and underinsured individuals. This company, a sister company of a business that markets medical travel for Canadians, promotes to uninsured and underinsured Americans affordable access to health care within eight U.S. states.

Using content analysis of company websites, this article is meant to provide an overview of Canada's medical travel industry. Content analysis of company websites reveals where these businesses are based, the health care destinations they promote to prospective clients, the health services they market, their core marketing messages, and whether companies market travel, hotel reservations, and tours in addition to advertising health services.

At present there is no resource that health researchers, clinicians, policy makers, journalists and other individuals can use as an accessible guide to analyzing both the general terrain of Canada's medical travel industry and the particular features of specific businesses. This article contributes to scholarship by putting a guide into the hands of health researchers and other parties interested in further exploring intellectual territory that is fascinating, complex, little-known, worthy of further study, and raises numerous ethical, social, legal, and public health concerns.

REFERENCES

1. Turner L: Medical tourism: Family medicine and international health-related travel. Canadian Family Physician. 2007, 53: 1639-1641.

2. Turner L: Canadian Medicare and the Global Health Care Bazaar. Policy Options. 2007, 73-77.

3. Turner L: "First World Health Care at Third World Prices": Globalization, Bioethics and Medical Tourism. BioSocieties. 2007, 2: 303-325.

4. Turner L: Quality in health care and globalization of health services: accreditation and regulatory oversight of medical tourism companies. Int J Qual Health Care. 2011, 23

5. Eggertson L: Wait-list weary Canadians seek treatment abroad. CMAJ. 2006, 174

6. Korcok M: Excess demand meets excess supply as referral companies link Canadian patients, US hospitals. CMAJ. 1997, 157: 767-770.

7. Katz S, Verrilli D, Barer M: Canadians' Use of U.S. Medical Services. Health Aff. 1998, 17: 225-235.

8. Katz S, Cardiff K, Pascali M, Barer M, Evans R: Phantoms In the Snow: Canadians' Use of Health Care Services in the United States. Health Aff. 2002, 21: 19-31.

9. Martin M: MS clinic's practices stir alarm. Winnipeg Free Press. 2011, Available from: http://www.winnipegfreepress.com/local/ms-clinics-practices -stir-alarm-113126744.html

10. Romaniuk R: Manitobans can head to Mayo Clinic. Winnipeg Sun. 2011, Available from: http://www.winnipegsun.com/2011/07/13/mayo-clinic

11. Brodbeck T: Medicare can't fight market. Winnipeg Sun. 2011, Available from: http://www.winnipegsun.com/2011/07/13/medicare-cant-fight-market

12. Industry Canada. Corporations Canada: Available at: https://www.ic.gc.ca/app/scr/cc/CorporationsCanada/fdrlCrpSrch.html;jsessionid=0000v9O3VPcCrmqiXhaZKPaOE2o:15dmaulj8?locale=en_CA

13. Floyd M, Izenberg D, Kelly B, et al: Medical services directory. Maclean's. 2006, Available from: http://www.macleans.ca/science/health/article.jsp?content=20060501_126220_126220

14. Find Medical Travel Services. Available at: http://www.findinghealthcare.ca/medical_tourism.html

15. Medical Tourism Facilitators. Tourism Review.com. 2009, 50-52. Available at: http://www.tourism-review.com/fm956/m6.pdf

16. India medical tourism destination 2009: healthcare without borders. The first ever medical tourism exhibition and conference in Canada [Internet]. 2009, Toronto: IMTD, Available from: http://www.imtd2009.com/

17. Pope C, Ziebland S, Mays N: Qualitative research in health care: Analysing qualitative data. BMJ. 2000, 320: 114-116.

18. Cormany D, Baloglu S: Medical travel facilitator websites: An exploratory study of web page contents and services offered to the prospective medical tourist. Tourism Management. 2011, 32: 709-716.

19. Lunt N, Hardey M, Mannion R: Nip, tuck and click: Medical tourism and the emergence of web-based health information. The Open Medical Informatics Journal. 2010, 4: 1-11.

20. Puzic S: Windsor police lay fraud charges against EcuMedical couple. Windsor Star. 2010, 2010: 2010-

21. Mason A, Wright KB: Framing medical tourism: An examination of appeal, risk, convalescence, accreditation, and interactivity in medical tourism web sites. Journal of Health Communication. 2011, 16 (2): 163-177.

22. Penney K, Snyder J, Crooks V, Johnston R: Risk communication and informed consent in the medical tourism industry: A thematic content analysis of Canadian broker websites. BMC Medical Ethics . 2011, 12: 17-10

23. York D: Medical Tourism: the Trend Toward Outsourcing Medical Procedures to Foreign Countries. Journal of Continuing Education in the Health Professions. 2008, 28 (2): 99-102.

24. Connell J: Medical tourism: Sea, sun, sand and...surgery. Tourism Management. 2006, 27: 1093-1100.

25. Sobo EJ: Medical travel: what it means, why it matters. Medical Anthropology. 2009, 28 (4): 326-335.

26. Sobo E, Herlihy E, Bicker M: Selling medical travel to US patient-consumers: the cultural appeal of website marketing messages. Anthropol Med. 2011, 18: 119-136.

27. Lunt N, Carrera P: Systematic review of websites for prospective medical tourists. Tourism Review. 2011, 66: 57-67.

28. Steinbrook R: Private Health Care in Canada. NEJM. 2006, 354: 1661-1664.

29. Gordon M, Berger P: The Alluring Myth of Private Medicine. CMAJ. 1996, 155: 404-406.

30. Charles C, Lomas J, Giacomini M: Medical necessity in Canadian health policy: four meanings and...a funeral?. Milbank Quarterly. 1997, 75: 365-394.

31. Flood C, Archibald T: The illegality of private health care in Canada. CMAJ. 2001, 164: 825-830.

32. Gray C: Visions of our health care future: Is a parallel private system the answer?. CMAJ. 1996, 154: 1084-1087.

33. Christou N, Efthimiou E: Bariatric surgery waiting times in Canada. Can J. Surg. 2009, 52: 229-234.

34. Legare J, Li D, Buth K: How established wait time benchmarks significantly underestimate total wait times for cardiac surgery. Can J. Cardiol. 2010, 26: E17-21.

35. Snider M, MacDonald S, Pototschnik R: Waiting times and patient perspectives for total hip and knee arthroplasty in rural and urban Ontario. Can J. Surg. 2005, 48: 355-360.

36. Lindberg M, Risk J: External Review of the Ontario Health Insurance Plan's Out-of-Country Program. 2007, 1-18. Available from: http://www.health.gov.on.ca/english/public/pub/ministry_reports/out_of_country_ohip/out_of_country_ohip.pdf

37. Korcok M: Ontario's move to limit out-of-province health care spending pays off in big way. CMAJ. 1993, 148: 425-426.

38. Morrow A: Man dies after controversial MS treatment, doctor says. The Globe and Mail. 2010, Available from:http://www.theglobeandmail.com/news/national/man-dies-after-controversial-ms-treatment-doctor-says/article1314585/

39. Alphonso C: Death of MS patient fuels debate over new treatment. The Globe and Mail. 2010, Available from:http://www.theglobeandmail.com/news/national/death-of-ms-patient-fuels-debate-over-new-treatment/article1807127/

40. McClure M: Woman with MS dies after treatment. Winnipeg Free Press. 2011, Available from:http://www.winnipegfreepress.com/canada/Woman-with-MS-dies-after-treatment--125269119.html

41. Puzic S: Company that linked Canadian patients to U.S. health care folds. The Windsor Star. 2010

42. Puzic S: Windsor police lay fraud charges against EcuMedical couple. The Windsor Star. 2010

43. Cohen T: New York medical brokerage gains ground in Canada. Times-Colonist. 2011

44. Kumar S, Breuing R, Chahal R: Globalization of Health Care Delivery in the United States through Medical Tourism. Journal of Health Communication. 2012, 17: 1-22.

45. Alleman BW, Luger T, Reisinger HS, Martin R, Horowitz MD, Cram P: Medical tourism services available to residents of the United States. J Gen Intern Med. 2010, 26: 492-497.

46. Shah KY: Current status and emerging trends of international medical outsourcing in the United States—a qualitative study. 2007, University of South Carolina doctoral dissertation, Archived at: http://www.webcitation.org/5x9wiGYUu

47. Crooks V, Snyder J: Medical tourism: What Canadian family physicians need to know. Canadian Family Physician. 2011, 57 (5): 527-529.

48. Snyder J, Crooks V, Adams K, Kingsbury P, Johnston R: The patient's physician one-step removed': the evolving roles of medical tourism facilitators. Journal of Medical ethics. 2011, 37 (9): 530-534.

49. Johnston R, Crooks V, Adams K, Snyder J, Kingsbury P: An Industry Perspective on Canadian Patients' Involvement in Medical Tourism: Implications for Public Health. BMC Public Health. 2011, 11: 416

50. Mehra N: Eroding Public Medicare: Lessons and Consequences of For-Profit Health Care Across Canada. Eroding Public Medicare: Lessons and Consequences of For-Profit Health Care Across Canada. 2008, 6: 1-169. Available from: http://www.web.net/ohc/Eroding%20Public%20 Medicare.pdf

51. Revah G, Bell C: Shopping for High-Technology Treatment in Another Province. Healthcare Policy. 2007, 2: 49-55.

52. Silversides A: Canada Health Act breaches are being ignored, pro-medicare groups charge. CMAJ. 2008, 179: 1112-1113.

53. Blackwell T: Is new MS research the real thing, or a media-driven frenzy. National Post. 2010, Available from:http://www.nationalpost.com/news/story.html?id=2475272

54. Weeks C: An uncertain future lies ahead for Zamboni's MS theory and its potential patients. The Globe and Mail. 2011, Available from: http://www.theglobeandmail.com/life/health-and-fitness/an-uncertain-future-lies-ahead-for-zambonis-ms-theory-and-its-potential-patients/article584981/

55. Dhalla I: Private Health Insurance: An International Overview and Considerations for Canada. Healthcare Quarterly. 2007, 10: 89-96.

Chapter 8

Honeymoon, medical treatment or big business? An analysis of the meanings of the term "reproductive tourism" in German and Israeli public media discourses

Sharon Bassan[1] and Merle A Michaelsen[2]

[1]Faculty of Law, Tel Aviv University, Ramat Aviv, PB 39040, Tel Aviv 69978, Israel.

[2]Department of Medical Ethics and History of Medicine, University Medical Center Göttingen, Humboldtallee 36, Göttingen 37073, Germany.

ABSTRACT

Background/Introduction

Infertile couples that travel to another country for reproductive treatment do not refer to themselves as "reproductive tourists". They might even be offended by this term. "Tourism" is a metaphor with hidden connotations. We will analyze these connotations in public media discourses on "reproductive tourism" in Israel and Germany. We chose to focus on these two countries since legal, ethical and religious restrictions give couples a similar motivation to travel for reproductive care, while the cultural backgrounds and conceptions of reproduction are different.

Results

Our research shows that the use of the metaphor "reproductive treatment" and its hidden messages depends on the writers' intention and the target population. Although the phenomenon of patients travelling for reproductive treatment can fit into the definitions of tourism, the term emphasizes aspects that do not reflect patients' reality. In both the German and the Israeli public media debate the term "reproductive tourism" is either used to criticize the economic aspects of the phenomenon or to attract patients as potential clients.

Conclusions

Ethicists should be cautious when borrowing metaphors like "reproductive tourism" from the public debate. Our findings support Penning's suggestion to use instead an unloaded term like cross-border reproductive care to describe the phenomenon in a more neutral way and to make it explicit whenever criticism is necessary.

KEYWORDS

Reproductive tourism; Metaphor; Medical tourism; Israel; Germany; Reproductive treatment; Infertility; Legal restrictions

1. BACKGROUND

In a study about so-called reproductive tourists the anthropologist Marcia Inhorn shows that many of the infertile couples going abroad for medical treatment do not regard themselves as tourists but instead as patients. They even feel offended by being called tourists because tourism is associated with fun, leisure or holidays, while they often experience their journey as highly stressful[1]. The term "reproductive tourism" is often used uncritically for a variety of practices. The article will describe the linguistic use and the dominant problems arising from it.

After introducing the basic facts of cross-border reproductive care and the general conditions that lead to the use of these procedures, we elaborate on the different legal-cultural backgrounds that motivate people from Israel and Germany to travel to other countries for reproductive treatment. We chose to focus on these two countries since legal, ethical and religious restrictions give couples a similar motivation to travel for reproductive care, while the cultural backgrounds and conceptions of reproduction are different.

Our aim is to examine if and how, by using the term "reproductive tourism", the public media discourse in Germany and Israel implicitly conveys certain value judgements. We will address reasons, if existing, for using a loaded term rather than a more neutral one, such as "cross-border reproductive care". By "neutral" we mean terms that are not morally loaded. The results of studies in the two countries will be compared, and hidden cultural, legal and social perceptions of the term "tourism" will be made explicit.

In light of the importance of the more general phenomenon of "tourism" in an era of globalization, our results contribute to a better understanding of the use of the term "reproductive tourism" and when and why it should be avoided.

1.1 The phenomenon behind the term

The terms "medical tourism" and "health tourism" are used to describe treatment or surgery which has been planned in advance to take place outside a patient's usual place of residence[2]. "Reproductive tourism" refers to infertile couples or individuals from one country purchasing reproductive or medical services from suppliers of such services in other countries, in order to conceive and have a child[3]. The term "reproductive tourism" covers various procedures which patients undergo such as *in-vitro* fertilization (IVF), egg and sperm donations, surrogacy, sex selection, and "pre-implantation genetic diagnosis" (PGD), whereas in other countries, the term also includes international adoption[4, 5, 6]. Other terms like “fertility tourism”, “procreative tourism” or “fertile tourism” all refer to the same practices.

According to the media[6, 7] and to professional journals[8], “medical tourism” and “reproductive tourism” have recently become a growing international industry due to cheap and fast means of travel, the availability of services, the amount of information on the internet, and the possibility to share medical practices[2].

So far there has not been any official registration or standardization of the collection of data of cross-border reproductive treatment. Therefore, only incomplete data are available. In 2009 about 3 – 5% of all medical treatments in the USA were carried out on non-U.S.-citizens, for example from Canada, Europe or Asia. In Europe approximately 7 - 10% of medical care is provided to foreign nationals, mainly from other European countries[9]. Shenfield et al. estimate that in Europe 11 000 – 14 000 patients are travelling for reproductive treatment annually, based on a survey conducted in 2008/2009[10]. The numbers of transactions in India, for example, show the explosive growth of this phenomenon: As Priti Sehgal states, in India “surrogacy is already an almost US$445 million business”, with a 150 percent rise in surrogacy cases during the last years[11].

1.2 Reasons to travel

Patients choose medical (and, as part of it reproductive) treatments abroad for different reasons: to avoid rationing, such as waiting periods in their countries of origin[12]; to benefit from the low costs or higher quality of services overseas; to secure their privacy[13], or to have access to services not provided in their home country. When national laws in affluent countries restrict or prohibit medical forms of procreation or citizens' access to fertility treatments, infertile individuals may privately exercise what they perceive to be their reproductive rights in other countries that have less restrictive laws regulating fertility transactions[14, 15, 16]. Such transactions mainly take place in the private market that is governed by the global economy, where the rules are mainly determined by liberal market values.

Countries that offer reproductive treatment not available in other countries therefore serve as a common destination for cross-border reproductive treatment. Today patients mainly travel from industrialized countries to less-developed nations when seeking affordable, high-quality treatment. For example, laws in the U.S. and UK do not allow a surrogate woman to charge the childless couple; therefore a smaller number of women tend to become surrogate mothers. In India there are no laws to prevent women from accepting compensation for surrogacy. Accordingly, it is easier for couples planning to have a baby to locate a surrogate mother in India than in the UK. In this light, there is a rising concern that women from vulnerable groups in target countries who often serve as surrogates or egg donors lack legal protection [17].

1.3 The legal and cultural background in Germany and Israel

The recent legislation on reproductive treatments in both Germany and Israel may foster patients travelling to obtain reproductive treatments elsewhere.

In Germany the Embryonenschutzgesetz (Embryo Protection Law, ESchG) of 1991 was the first legislation to regulate artificial reproductive technologies. During the preparation of this law, the German lawyer and former president of the Federal Constitutional Court, Ernst Benda, called reproductive treatment that involves "unnatural splitting of motherhood" a threat to human dignity[18]. The inviolability of human dignity is established in the first paragraph of the German constitution. The fact that Benda referred to human dignity underlines the political and social importance of this issue in Germany.

Surrogacy and egg donation are forbidden by law in Germany. Other treatments, like IVF treatment only became available for people in a heterosexual marital or marital-like relationship; homosexual couples still are excluded from reproductive treatment.

Therefore, homosexual couples and infertile couples who choose egg donation or surrogacy have no other option but to look for such services elsewhere. Germans, for example, often travel to Belgium or Spain where more liberal legislation allows for egg donation and where homosexual couples also have access to reproductive treatment [19]. In German public media, the magazine "Der Spiegel" describes the role of the law as creating a global market for reproductive treatment: "Eventually, globalization brings dozens of new businesses into the country. For some time, strict laws in Europe and elsewhere have led to a flourishing of fertility tourism. The weak dollar has caused a veritable baby boom: some U.S. agencies are doing more than 40 percent of their business with customers from overseas[a]" [20].

In contrast, in Israel a more liberal choice policy is ensured by legislation. Parenthood has a broad social value and is promoted by a range of different cultures within the country, most notably Jewish culture and its reliance on religious scripture[b]. In addition, the Holocaust and the recent history of the state of Israel have created collective a pro-natalist awareness [21]. The trauma of the Holocaust and a communitarian sentiment to increase the number of Jewish people in Israel have been recognized as contributory as motivation for the liberal regulation of artificial reproductive technologies in the country [22]. This attitude towards reproduction in Israel may be important in explaining why travel to receive reproductive treatment is regarded as a means to a justifiable end – parenthood.

In Israel, every citizen is entitled to "infertility diagnosis and therapy". Artificial fertilization "for the purpose of bearing a first and second child - for couples who do not have children from their current marriage, and also for a childless woman who wishes to establish a single parent family", as specified in an addendum to the law, is part of the basic health package of medical services and medicines[c]. Historically, the Clalit health fund had no limitations on the number of treatments for the insured.

Most procedures conducted beyond the country's borders are egg donation and surrogacy. Until recently (June 2010), when a new egg donation law was passed, it was not legally possible to receive an egg donation from a woman who was undergoing the process for the sole purpose of donating her eggs. Only women who underwent IVF treatment for themselves were allowed to donate eggs, but most of these women chose to freeze their eggs for later and avoid repeating the complicated procedure of egg retrieval. Therefore, in Israel the demand for eggs exceeded the supply.

With the new egg donation law introduced in June 2010, it is now possible to recruit egg donation from women who are not undergoing the procedure for their own purpose, but rather in exchange for compensation [23]. For now, private clinics still offer the possibility of fertilization from donors' eggs in medical institutions outside Israel (e.g. from Cyprus, Ukraine, and Romania). According to a media report in the television's weekend magazine (February 27th 2004), couples travel to places where eggs are retrieved from a paid donor, and are transplanted into the womb of the recipient. After this legislative change, the future of cross-border reproductive care for egg donation depends on the development of supply.

Legal restrictions on surrogacy for same sex couples and single parents led numerous couples to consider surrogacy outside of Israel (e.g., India). With no known intention to change this restriction, it is unlikely that travelling abroad for surrogacy, at least for same sex couples, will decrease or stop in the near future.

In conclusion, in Germany assisted reproductive treatment is regulated by restrictive legislation based on moral concerns, while in Israel there is a more liberal attitude toward options of reproductive technology. However, in both countries homosexual couples and single parents do not have access to all assisted reproductive treatments available. Thus, in Israel and Germany, policies create a demand for egg donation and surrogacy abroad and have engendered public debate about "reproductive tourism".

2. METHODS

To identify different use of the term "tourism", we conducted an internet-based pre-search by keywords. We then analyzed identified articles to determine the precise terms used to describe the practice and phenomenon of travelling patients[d]. In order to define how the public medica discourse in Germany and Israel make use of the term "tourism", different contexts and connotations were analyzed through the following questions:

1 Are the different terms used in distinct contexts?

2 What characteristics of the practice and phenomenon are considered when the term "tourism" is used in German and Israeli public media?

3 What characteristics are considered when the term "tourism" is avoided?

We expected that cultural connotations would be easier to detect when concentrating on discourses in the respective native languages. We therefore focused on websites such as I-forums, online newspapers, articles, etc. in both countries, and in Germany focused upon printed media as well. Through the analysis we identified particular cultural, legal and social perceptions of the use of the term "tourism". We compared the results from Israel and Germany and suggested reasons for identified differences.

2.1 Results and analysis

In our analysis of the public media discourse we will discern different uses of the term "reproductive tourism". We will, in particular, focus on the metaphorical use of the term. Metaphors are a linguistic instrument to implicitly express values as well as facts. The public media discourse reflects and also influences public opinion; it thereby can have a normative function. Consequently, analysis of the different uses of the term can show how a situation or phenomenon is described and evaluated in a society. We herein present the main findings regarding the contexts and references of the analyzed terms; we classified the sources, background conditions, and target population of the references, as well as the different contexts according to salient common characteristics.

2.2 The public media discourse in Israel

In Israel, we searched the internet in Hebrew for different references that use any term related to tourism in the field of reproductive treatment. About 30 items were found, not all of which were relevant. In all, 16 relevant references were identified and analyzed. The characteristics were considered according to the context and the different use of the term. The Israeli public is very interested in reproductive technologies. Search results from the internet clearly showed that all of the popular newspapers deal with subjects relating to cross-border reproductive care. It should be mentioned that articles also used other words when reviewing the phenomenon. The subject was also raised in the parliament; therefore two references were published by the ministry of health and the parliament's research centre. Other references were retrieved from informative/publication websites of clinics, supplying information regarding their services and the process to potential clients, or from patients forums.

In Israel, the terms “fertility tourism” or “reproductive tourism” ("תיירות פריון") are used to imply the consequence of regulation differences between countries and in order to lobby for a change in legislation that would make these “services” legally accessible[24]. The language reflects a market model referring to a “deficit” in eggs: “The race after the eggs: the rising demand in IVF brought on a flourishing global market of eggs”[25]. Yet, “egg tourism” was also used in the Knesset, when discussing the new egg donation bill as something not wanted in Israel[26].

When the right to parenthood is the focus of the article, a more neutral term is used (“cross-border reproductive care”)[27]. When patients in Israel are directly addressed, the term “tourism” is never used ("תיירות"), but instead is referred to as “to purchase eggs” ("לרכוש ביציות"), “to import” ("לייבא") or simply “egg donation abroad” ("לייבא")[16, 28, 29, 30]. However, we have found patients using the term “tourism” in I-forums, for example, when talking about a “travel agent” that “specializes” in “fertility tourism”. The forums are often used by infertile people seeking information about an opportunity to have a child. Other prospected parents receiving information about the procedure in a surrogacy centre stated: “...it runs like *a business, a baby market*, they talk with us about the *costs of a womb for rent*”[31].

The expression “womb for rent” (בחו"ל""תרומת ביציות) was used in several contexts: First, by feminist organizations that objected to the law that enables surrogacy in Israel: “the law encourages treating women as an instrument to deliver babies – a womb for rent”[32]. Second, by an infertile woman who described that she wanted minimal social contact with the surrogate mother: “I only wanted a womb for rent”[33]. Third, when describing negative consequences of poverty: “The solution to economic agony: a womb for rent”[34]. Another expression related to market vocabulary was “surrogacy industry” ("רחם להשכרה"); this was used in an article about surrogacy in India to express human rights activists’ objections. Another article dealing with the “international fertility market” in India also used the expression “surrogacy industry”[35]. Another example of market vocabulary is the expression “to buy eggs” ("תעשיית הפונדקאות") that was cynically used in Israel in order to describe senior gynecologists offering infertile women the opportunity “to buy” eggs abroad, while choosing the “donor” according to her physical traits[36].

In summer 2009, the state of Romania charged some Israeli doctors with trafficking human eggs. The doctors arrived in Romania with their Israeli patients in order to recruit eggs from Romanian women. Most media reports dealing with this affair used terms like "egg trafficking"[37], or "egg sale"[38]. Since the incidence in Romania, a new legal vocabulary that implies trade and trafficking, was developed to express concerns regarding the ramifications of such practices.

It is interesting that the terminology in Israel appears to be quite functional. Different terms are used for various target populations, and for different purposes: when recruiting potential clients (or customers), and/or when addressing infertile couples, it is customary not to use the term "tourism". Rather, a more neutral, descriptive term will be used, such as "IVF patient/service". Moreover, when addressing these populations, the term "donation" is used to describe the supply of reproductive services despite the fact that the intention is to buy and sell these services. Nevertheless, the term "tourism", or even terms such as "trafficking" that may have a more negative connotation, are used when discussing the economic aspects of such practices, or as legal accusations against doctors.

2.3 The public media discourse in Germany

In Germany, we reviewed five of the most important and opinion-leading newspapers and magazines: "Der Spiegel", "Die Zeit", "die tageszeitung", "Frankfurter Rundschau", "Frankfurter Allgemeine Zeitung (FAZ)"[e].

Altogether 63 articles published in the years 2000 - 2010 were found, 35 used a tourism metaphor or treated the topics of egg donation or surrogacy in the context of travelling for reproductive health care; 18 out of the 35 articles analyzed used a term that contains the word "tourism" in the context of medical treatment. Three (3) articles of the 18 used a term related to "tourism" generically to indicate the phenomenon. Thirteen (13) articles out of 18 mainly dealt with the commercial aspects of the phenomenon. The FAZ article "500.000 eingefrorene Embryos"("500.000 frozen embryos") states: "The business of IVF, sperm banks, egg cell donation and surrogate mothers has become one of the fastest growing industries of South California[39]. "Only 9 of 35 articles mentioned the donors' point of view. Of these 9 articles only one, reporting on the surrogate mother and/or egg donators, used a term related to "tourism", the other 8 did not make any reference to the term. Six (6) of these articles dealt with commercial aspects of surrogacy, but included the perspective of the infertile couples. For example, the article "Verbotene Kinder"

("Forbidden Children") appearing in "Die Zeit" elaborates in detail on the stresses of the infertile couple as well as the surrogate mothers' situation[40]. In the article, vocabulary from the market is used to describe the economic aspects, yet the term "tourism" is avoided. In the whole sample, the physicians' point of view is mentioned more often than the donors' perspective: Twelve (12) of the articles report on physicians' perspective, six (6) of them using the term "tourism".

In the public media discourse in Germany, the terms "fertility tourism" or "reproductive tourism" are most commonly used in a commercial context, for example, by calling/denouncing the practice of travelling for reproductive purposes an industry, a business, or an opportunity for making money. The German magazine "Der Spiegel", for example, writes: "Those to whom the *supply* of the German baby mills is not sufficient, turn to foreign countries because the test-tube procreation *market* has become a global one for a long time"[41]. Another "Der Spiegel" article, "Babies made to order" states: "A child at your request? No problem in the USA: There are companies finding surrogate mothers, egg or sperm donation in California. The business booms – also with clients from overseas"[20]. The German newspaper "Die Zeit" states: "Nowhere else in the world is it allowed to do *business* with sperm and egg cells according to the *rules of supply and demand*"[42].

The term "tourism" in the German public debate emphasizes the business aspect of the phenomenon. When the physicians' point of view is reflected, the use of the metaphor is ambivalent. When it is avoided, the article usually discusses the phenomenon in a broader perspective such as that of the person travelling abroad for these services and/or the one of the surrogate/egg donor. We find that the use of the term "tourism" is not a coincidence. It is interesting that both in Israel and Germany, the term was not used to express the fact that the travelling is done for a specific purpose, for pleasure, or cultural enjoyment. Rather, it is used to illustrate, and to criticize the economic aspects of these services and practices.

2.4 Tourism and reproductive treatment

One can identify two main contexts, in which the term tourism is usually understood and interpreted: First, the context of leisure and fun, and second, the industrial and economic context. According to common understanding, "tourism" refers to travel to another country for a vacation, in order to enjoy its natural and/or cultural resources.

But the term "tourism" is also an economic term. The Oxford Dictionary defines tourism as "the commercial organization and operation of holidays"[43]. Tourism as part of the service industry is of high significance for global economics.

The World Tourism Organization defines tourists as people who "travel to and stay in places outside their usual environment for more than twenty-four (24) hours and not more than one consecutive year for leisure, business and other purposes not related to the exercise of an activity remunerated from within the place visited"[44].

This definition does not limit the purpose of tourism, and we think that this definition can easily fit "reproductive tourism" in three dimensions. First, it is well known that there are many specific market segments that use tourism for "other purposes", like sports tourism, music tourism, business tourism etc. Receiving reproductive health care can be seen as such a purpose. Describing travellers as "fertility tourists" also matches the concept of tourists travelling for business purposes or in search of a cure. Second, "the purpose of leisure" can express the leisure of having a family or enjoying the idea of having a child. Finally, even the classical motive "to enjoy a country's culture" can apply to medical tourists. They benefit from a country's more liberal attitude(s) towards reproductive treatment, and possibly from economic circumstances that motivate women to supply reproductive services to anyone who can afford to pay.

2.5 Ethical analysis

Although the phenomenon of patients travelling for reproductive treatment can fit into definitions of "tourism", some ethicists have raised concerns regarding the implicit normative connotations of the term. Ritu Priya, at the Centre for Social Medicine and Community Health at Jawaharlal Nehru University in New Delhi, remarks that using the notion "health tourism is likely to distort the perspectives of health providers, promoting medicine as a purely commercial venture" [45]. The legal scholar Richard Storrow also criticizes the use of the term by providers; he argues that marketing the reproductive treatment industry implies that the journey is not only framed by the "fantasy" of conceiving a child, but also to do so during romantic holidays. An exotic vacation is promised, during which the medical treatment is accompanied by entertainment and pleasure, instead of staying in a hospital and receiving medical treatment with all its potential side-effects burdens and risks. This concept is presented as analogous to planning a vacation with the help of a travel agent, rather than scheduling a medical procedure [46]. By using the term "tourism" that, according to Storrow, is mainly understood "as a type of travelling that involves leisure, pleasure and free time", the providers may conceal the realities of the journey, i.e., the medical treatment.

Yet, given that infertile couples often describe their condition as devastating, and their effort to conceive a child as stressful and requiring enormous physical and emotional strength, it seems difficult to harmonize the realities of "reproductive tourism" with the idea of tourism as pleasure travel [5]. In her empirical research with couples that travelled to other countries for reproductive treatment, Marcia Inhorn discovered that, for the patients, the use of the term seems insensitive given the stressful and expensive journey they had to undertake. Most of the patients would prefer legal, trustworthy, and economical services available closer to their home(s), having the impression that "...'reproductive tourism' sounds like a 'gimmick', which does not take infertile people's suffering seriously"[47].

Therefore, different terms have been suggested. Roberto Matorras, president of the "Sociedad Espanola de fertilidad", argues that the term "tourism" may be offensive both to couples and to professionals involved in the process [48]. In order to focus on the patients, Matorras suggests the term "reproductive exile". In his opinion, this better expresses the difficulties and constraints faced by infertile patients who are forced to travel globally for assisted reproduction.

Guido Pennings agrees with Matorras that "tourism" is a derogatory term when used in the medical context. The term, he argues, suggests that the desire of the patients is insignificant, or that they are looking for something exceptional or "bizarre", although most patients simply want to avoid long waiting lists or high costs. Pennings, however, holds that the term "exile" creates the misleading impression that patients are forced to go abroad as a punishment. In a reply to Matorras, he suggests replacing the term "tourism" by a more neutral term, such as "cross border reproductive care", which does not carry the problematic connotations of "tourism", and its accompanying value judgments [49]. Contrary to this criticism, Kristen Smith states that the term "tourism" in a medical context *can* be used in a constructive way, as "...the bioethical dimensions of the practice and industry are brought into a much sharper focus than could be achieved with a more benign descriptor such as the term 'medical travel'"[50].

Our findings show that in both Germany and Israel, the use of the metaphor "reproductive tourism" and its hidden messages depend on the writers' intention and the target population. We can differentiate two opposite positions: the one of the critics, and the one of the providers. While the critics' intention in using the term is to illustrate their criticism by using the metaphor "tourism" as an economic term, providers use it to justify their business on one hand, and to trivialize the efforts of their potential patients on the other. Providers and the public media express different values of the phenomenon by using the same term. Both can make use of the term according to agendas that either condemn or trivialize the underlying social phenomenon.

3. CONCLUSIONS

In the public debate the term "reproductive tourism" is used to highlight certain aspects of seeking medical treatment to conceive or bear a child. Metaphors are common linguistic instruments illustrate a particular position. But metaphors usually do not reflect all aspects of the phenomenon. "Reproductive tourism" does not reflect patients' reality, in particular the stress and exertion involved. Patients therefore feel offended by the term. Thus, some ethicists suggest using different metaphors for the practice, and its various aspects and actors.

Terms like "tourism" or "exile" tend to be biased in that they have represent value judgements, that implicitly stress a certain aspect or meaning. In an ethical debate, however, positions and arguments should be explicitly and clearly stated. Implicit criticisms through the use of metaphors lead to misunderstandings, imprecision and inappropriate generalization. Therefore, ethicists should be cautious when using these metaphors that are adapted from public discourse. Our findings support Guido Penning's suggestion to use unloaded term, such as cross-border reproductive care to denominate the practice in a more neutral way, and to ensure that its meanings is explicit, when engaging in debate and/or criticism.

ENDNOTES

[a]All translations are made by the authors.

[b]Orthodox Jews consider it to be a religious duty to “be fruitful and multiply”. The curse of female barrenness is a consistent and profound theme in the bible and in Jewish tradition. The words of Rachel to Jacob “give me children or else I die” (Genesis 30:1) continue to reverberate in Israeli culture. According to *halakha* (Jewish law), which governs the laws of marriage and divorce for Jews in Israel, a man has the right to divorce his wife if the marriage has failed to produce children over ten years.

[c]Paragraph 6(e) of the Second Addendum to the National Health Insurance Law, 1994, *S.H.* no. 1469, p. 183.

[d]Four terms that use the tourism metaphor were found: “Fortpflanzungstourismus” (“reproduction tourism”), “Fertilitätstourismus” (“fertility tourism”), “Reproduktionstourismus” (“reproduction tourism”) and “Fruchtbarkeitstourismus” (“fecundity tourism”), in Hebrew: "תיירות ביציות" "תיירות פריון", "תיירות פוריות", "לקנות ביציות". We searched all articles that use one of these “tourism” terms. Time period for our search was in both contexts 10 years: (Jan.01st, 2000 until Jun.19th, 2010).

We also used other keywords “Leihmutter” (“surrogate mother”), "אם פונדקאית" "תיירות פונדקאות" and “Eizellspende” (“egg donation”), "תרומת ביציות" linked to the words “Ausland” (“foreign country”), "חו"ל" or “Reise” (“travel”), "נסיעה" in order to find all articles dealing with our research topic.

[e]“Der Spiegel” is known for its investigative journalism, “die tageszeitung” for its left-wing tendencies. “Die Zeit” is an independent weekly newspaper that is widely recognized as liberal. While the “Frankfurter Rundschau” is considered as social liberal the “Frankfurter Allgemeine Zeitung” is known as classical liberal with conservative tendencies.

ACKNOWLEDGEMENTS

The authors would like to thank all colleagues who contributed to this manuscript. It is a pleasure to thank Silke Schicktanz and Aviad Raz who initiated this collaboration with the workshop "Winter school: the culture and ethics in biomedicine" in 2009. We thank Swantje Michaelsen and Sabina Fazli for the English correction of the manuscript. Our special gratitude goes to Claudia Wiesemann for the supervision, inspiration and support she extended.

AUTHORS' CONTRIBUTIONS

SB participated in the design of the study, performed the analysis especially for the parts of the article that concern the situation in Israel and drafted the manuscript. MAM participated in the design of the study, performed the analysis especially for the parts of the article that concern the situation in Germany and contributed in drafting the manuscript. All authors read and approved the final manuscript.

REFERENCES

1. Inhorn MC, Shrivastav P: Globalization and reproductive tourism in the United Arab Emirates. Asia Pac J Public Health. 2010, 22: 68S-74S.
2. Mecir A, Greider K: Traveling for treatment - soaring U.S. Health costs are driving more Americans abroad for medical treatment. AARP Bull. 2007
3. Howze KS: Medical tourism - symptom or cure?. Ga L Rev. 2007, 41: 1013-1014.
4. Pennings G: Reproductive tourism as a moral pluralism in motion. J Med Ethics. 2002, 28: 337-341.
5. Storrow RF: Quests for conception: fertility tourists, globalization and feminist legal theory. Hastings Law J. 2006, 57 (295): 295-330.
6. Cross J: Reproductive tourism: traveling abroad for IVF.http://www.healthnews.com/family-health/pregnancy-childbirth-parenting/reproductive-tourism-traveling-abroad-ivf-1484.html,
7. DeMicco FJ: Medical tourism growing worldwide. 2004

8. Rupa C: Trade in health services. 2002, 158-: Bulletin of the WHO
9. Adamson D: Global cultural and socioeconomic factors that influence access to assisted reproducitve technologies. Women's Health. 2009, 5: 351-355.
10. Shenfield F, De Mouzon J, Pennings G, Ferraretti AP, Andersen AN, De Wert G, Goossens V: Cross border reproductive care in six European countries. Hum Reprod. 2010, 25: 1361-1368.
11. Sehgal P: The wip. Reproductive tourism soars in India: adoption and surrogacy laws have Yet to catch Up. 2008,http://www.thewip.net/contributors/2008/10/reproductive_tourism_soars_in.html,
12. Terry N: A flat world: medical tourism and outsourcing. New Eng L Rev. 2007, 433-434.
13. Fried BJ, Harris DM: Managing healthcare services in the global marketplace. Front Heal Serv Manag. 2007, 24: 3-18.
14. Anderson ES: Is Women's labor a commodity?. Phil and Pub Aff. 1990, 19: 71-75.
15. Held V: Rights and goods: justifying social action. 1989, Chicago: Univ. of Chicago Press
16. Barken N, El Hai L, Beno G, Turgeman M, Eichner I, Brut T: All is Gold: on the egg industry for Israelis abroad.http://www.ynet.co.il/articles/0, 7340, L-3749827
17. Palattiyil G, Blyth E, Sidhva D, Balakrishnan G: Globalization and cross-border reproductive services: ethical implications of surrogacy in India for social work. Int Soc Work. 2010, 53: 686-700.
18. Ziegler U: Präimplantationsdiagnostik in England und Deutschland: ethische, rechtliche und praktische Probleme. 2004, Frankfurt, New York: Campus
19. O'Connell Davidson J: The rights and wrongs of prostitution. Hypatia. 2002, 17: 84-92.
20. Hornig F: Baby auf Bestellung. Der Spiegel. 2008, 52-57.

21. Shalev C, Gooldin S: The uses and misuses of *In vitro* fertilization in Israel: some sociological and ethical considerations. Nashim: A J Jewish Women's Stud & Gender Issues. 2006, 12: 151-176.

22. Shapira-Rosenberg E: Make children with love (or don't make them at all). http://www.ynet.co.il/articles/0, 7340, L-3749827, 00.html,

23. Shalev C, Werner-Felmayer G: Patterns of globalized reproduction: Egg cells regulation in Israel and Austria. Isr J of Health Policy Res. 2012, 1: 15-10.1186/2045-4015-1-15.

24. Ministry of law: Comparative research about surrogacy.http://www.health.gov.il/Download/pages/insler4_1.pdf,

25. Shachaf T: The race after the eggs. 2004,http://www.nrg.co.il/online/16/ART/777/234.html,

26. Sub-committee of labour, welfare and health: Following the egg donation bill: Protocol No.3.http://www.knesset.gov.il/protocols/data/rtf/avoda/2008-06-24-04.rtf,

27. Oren R: Egg donation as the right to Be a mother in Israel.http://www.news1.co.il/Archive/003-D-39957-00.html?tag=16-13-16,

28. Gal I: On reproduction, IVF, and what's between. http://www.ynet.co.il/articles/0, 7340, L-3220514, 00.html,

29. Zeidman D: Mamy. Egg donation - How is it done?. 2006, http://mamy.co.il/article.aspx?IDP = 2261,

30. Rofen A: Egg donation abroad.http://www.rofenashim.co.il/default.asp?kat=149,

31. Blikov SM: NRG. Parents meeting: more couples in Israel choose to have a baby with a surrogate. 2009,http://www.nrg.co.il/online/29/ART1/898/954.html,

32. Chayut I: A womb for rent or a reasonable solution?.http://www.haaretz.co.il/hasite/pages/ShArtPE.jhtml?itemNo=88273&contrassID=2&subContrassID=11&sbSubContrassID=0,

33. Negev E: A womb for rent. http://www.ynet.co.il/Ext/Comp/ArticleLayout/CdaArticlePrintPreview/1, 2506, L-2431954, 00.html,

34. Yagne Y: The solution to economic agony: a womb for rent.http://www.haaretz.co.il/hasite/spages/1072189.html,

35. Dresler T: India: A womb for rent and a first hand kidney in bargain price. http://www.ynet.co.il/articles/1, 7340, L-3362458, 00.html,

36. Levi V: Bring six, bring seven, bring eight children. Haaretz online. 2005

37. Hosu M, Even D, Yagna Y, Eyadat F, Ravid B: Romania to charge Israeli doctors for trafficking human eggs.http://www.haaretz.com/hasen/spages/1102054.html,

38. Sinai R: Women against the egg law: hurting the donors.

39. Hupertz H: 500 000 eingefrorene Embryos. Frankfurter Allgemeine Zeitung. 2006

40. Spiewak M: Die Zeit. Verbotene Kinder. 2010,http://pdf.zeit.de/2010/17/Leihmutterschaft.pdf,

41. Blech J, Lakotta B, Noack H: Babys auf Rezept. Der Spiegel. 2002, 70-80.

42. Böhm A: Im Land der Kindermacher. Die Zeit. 2001

43. Oxford dictionary of English: 2nd editionOxford, New York: Oxford University Press; 2003.

44. UN World tourism organization: 2nd Technical manual: collection of tourism expenditure statistics; 1995.

45. Mudur G: Hospitals in India woo foreign patients. BMJ. 2004, 328: 1338

46. Burkett L: Medical tourism - concerns, benifits, and the American legal perspective. J Legal Med. 2007: 223-226.

47. Inhorn MC, Patrizio P: Rethinking reproductive "tourism" as reproductive "exile". Fertil Steril. 2009, 92: 904-906.

48. Matorras R: Reproductive exile versus reproductive tourism. Hum Reprod. 2005, 20: 3571

49. Pennings G: Reply: reproductive exile versus reproductive tourism. Hum Reprod. 2005, 20: 3571-3573.

50. Smith K: The problematization of medical tourism: a critique of neoliberalism. Dev World Bioethics. 2012, 12: 1-8.

Chapter 9

Medical tourism in india: Perceptions of physicians in tertiary care hospitals

Imrana Qadeer and Sunita Reddy

Center of Social Medicine and Community Health, School of Social Sciences, Jawaharlal Nehru University, New Delhi 110067, India

ABSTRACT

Senior physicians of modern medicine in India play a key role in shaping policies and public opinion and institutional management. This paper explores their perceptions of medical tourism (MT) within India which is a complex process involving international demands and policy shifts from service to commercialisation of health care for trade, gross domestic profit, and foreign exchange. Through interviews of 91 physicians in tertiary care hospitals in three cities of India, this paper explores four areas of concern: their understanding of MT, their views of the hospitals they work in, perceptions of the value and place of MT in their hospital and their views on the implications of MT for medical care in the country. An overwhelming majority (90%) of physicians in the private tertiary sector and 74.3 percent in the public tertiary sector see huge scope for MT in the private tertiary sector in India. The private tertiary sector physicians were concerned about their patients alone and felt that health of the poor was the responsibility of the state. The public tertiary sector physicians' however, were sensitive to the problems of the common man and felt responsible. Even though the glamour of hi-tech associated with MT dazzled them, only 35.8 percent wanted MT in their hospitals and a total of 56 percent of them said MT cannot be a public sector priority. 10 percent in the private sector expressed reservations towards MT while the rest demanded state subsidies for MT. The disconnect between their concern for the common man and professionals views on MT was due to the lack of appreciation of the continuum between commercialisation, the denial of resources to public hospitals and shift of subsidies to the private sector. The paper highlights the differences and similarities in the perceptions and context of the two sets of physicians, presents evidence, that questions the support for MT and finally analyzes some key implications of MT on Indian health services, ethical issues emerging out of that and the need for understanding the linkages between public and private sectors for a more effective intervention for an equitable medical care policy.

BIOGRAPHIC SKETCH

Imrana qadeer

Imrana Qadeer's areas of research interests are organization of health services, Primary Health Care, Political Economy of Health, Epidemiology, Health Policy Analysis, Research Methods, Systems Research, Women's Health and Nutrition. She has published her research in national and international journals and books. Her forthcoming book is Health of the Dalit Women in India.

As a Professor at the Centre of Social medicine and Community Health, School of Social Science, Jawaharlal Nehru University, New Delhi she taught and guided research in public health for 34 years. Over 2008-2010, she was offered the prestigious J P Naik Senior fellowship by the Centre for Women's Development Studies, New Delhi. An MD in Pediatrics and a clinical registrar at the All India Institute of Medical Science, New Delhi she moved into public health to develop a critique of the bio-medical approach to public health in India. She worked with several formal institutions such as, the Planning Commission, the Population commission, and the monitoring Committee of the National Rural Health Mission. She was a member of the Standing Committee, currently a visiting Prof. at Council for Social Development of the University Grants Commission on Women's Studies Centres in Indian Universities and is also associated with several grass root organizations working on health issues.

Sunita Reddy

Dr. Sunita Reddy is an Anthropologist, specialized in medical anthropology, currently teaching in the 'Center of Social Medicine and Community Health', School of Social Sciences, Jawaharlal Nehru University. Dr. Reddy has been teaching and researching on Public Health issues. Her areas of research are women and children's health, public private partnership in health services, Disaster studies and Medical Tourism. She has done a longitudinal research on Rehabilitation Post Tsunami in Andaman and Nicobar Islands and published a book 'Clash of Waves' by Indos Publishers, New Delhi (2013). She has published research papers widely in peer reviewed journals and has presented papers in many national and International conferences. She has been delivering lectures for Academic Staff colleges, for Indian Foreign Service probationers and other institutions. She is a Core group member, in formulation of guidelines on psycho-social health, widows and children in disasters, for National Disaster Management Authority (NDMA) and also member in expert group for "CBRN Disaster Management: A Step Towards Capacity Building & Resilience in Industries" throughout the country.

1. INTRODUCTION

Medical Tourism in Asia is estimated as worth $ 4 billion by the year 2012 [1]. The critical factors that led to cross border travel to seek superior medical care are the rising cost of care in the developed world and presence of uninsured or underinsured people there. Patients from developed world are now seeking care in countries where hi-tech is available and inexpensive. Huge cost differentials (Asian countries charging 20% of the cost of US and UK), cheaper airfare, 'world class' hi-tech medical care today offers business opportunities for corporate hospitals in Asian countries.

Medical Tourism (MT) in India can be traced to the increasing numbers of corporate hospitals in the metropolitan cities. With increasing liberalization since the 80s, tertiary care in India's health sector was opened to private sector provisioning and public institutions permitted to accept private investments in the hope of enhancing revenues [2]. The Tenth and Eleventh Plans [2, 3] welcomed MT, commercialisation of the public sector and a hi-tech tertiary sector market as a part of reforms. The drafts of the National Health Bill [4] and the 12th Five year Plan of India emphasised the role of private sector in medical care [5]. This privatization and commercialization that transformed medical care from a service to a commodity was a conscious policy decision accompanied by a range of subsidies including land, equipment and drug imports etc. [2, 3, 6]. It also attracted medical experts from the public hospitals who had received state supported medical education [7]. Like any other business in welfare services in India, medical industry also started attracting clients through capturing the high-tech services. The health services were transformed from a national medical priority based network of institutions to one that offered the services- elective (could be planned in advance) and tuned to the needs of the global patients.

Thus, the emergence of medical tourism within India's policy frame at the turn of the century cannot be seen as an outcome of international influences alone. The ideological underpinnings of the Indian State itself are reflected in the policy shifts towards commercialisation and growth of MT that transforms health services into a source of trade, gross domestic profit, and foreign xchange [8]. The government policy of merging medical expansion and tourism was announced by Finance minister in his 2003 annual budget speech, where he called for 'India becoming a global health destination' [9]. Thus chains of such institutions and medi-cities got underway. India's XIth Plan mentions a list of corporate hospitals, which provide high-end health care services through business process outsourcing [3], p.275 and the XII Plan depends heavily on the public private partnerships in financing of tertiary care as a way to handle the financial challenge [5]. Confederation of Indian Industry (CII) and Indian Health Care Federation (IHCF) wants to establish an Indian health care brand synonymous with 'safety, trust, and excellence' [10]. The good public sector tertiary hospitals too are being geared to practice MT. The assumption behind these policies is that the revenue thus, generated will add to the economic growth that is the basis of welfare. The fast growth of private tertiary sector hospitals and their use indicates profit maximisation for medical business and need satisfaction for the upper middle classes, but what of the physicians among them trained in the science of medicine and public health? Are they satisfied and aware of emerging ethical contradictions? These are issues this paper helps us answer.

The paper focuses on the perceptions of MT among senior physicians around four areas of concern: their understanding of MT, their perceptions and views on the hospitals they work in, the value and place of MT in their hospitals and the implications of MT for medical care in the country. The validity of their perceptions is then examined against available literature on the subject. By contrasting these views of public and corporate tertiary care physicians the paper also highlights their contrasting ethics of medical care.

2. METHODS

This study focuses on the perceptions of MT among senior physicians from selected public and corporate tertiary hospitals. Though based on a relatively small sample of 91 physicians, it is critical as these physicians influence the health policies through their influence on public opinion specially the middle class, their advisory and consultancy roles in health planning at the Ministry of Health and Family Welfare of the Centre and State Governments, and micromanagement of their departments and institutions. The study is based on in-depth interviews as its main tool along with observations, hospital records and secondary literature. Open-ended interview schedules shown in Additional file 1 and probes in Additional file 2, wherever necessary were used. The interviews on an average took one hour and the responses were noted verbatim.

Two reputed public and two well known corporate hospitals in each of these three cities were selected purposively. Till 2010, there was no Ethics Review Board constituted for social science research either at the School of Social Sciences or at the Jawaharlal Nehru University Level. Following the general ethical guidelines practiced in academics the authors took written consent from the hospital authorities and verbal consent from the physicians to interview and publish their views. Anonymity of both- the physicians and the hospitals is maintained. Physicians from four public and six corporate tertiary hospitals from three metropolitan cities - Delhi, Chennai and Hyderabad - were selected. In Chennai, public hospitals refused permission due to some problems with the media just before the fieldwork in 2006-2007. The departments of cardiology, orthopaedics, surgery, and paediatrics with high focus on MT, other departments with or without MT and medical administrators were also chosen for interviews. The selection of physicians was thus purposive depending on their willingness for the interview. While writing, the method of paraphrasing multiple quotes was used both for bringing out commonalities and differences between the physicians of public and corporate tertiary sectors.

Table 1 gives the distribution of physicians in the study sample. The interviews from the corporate hospitals in three cities covered 52 physicians and from public sector hospitals in two cities covered 39 physicians.

3. FINDINGS

3.1 Understanding of medical tourism

Physicians in the public and corporate sectors viewed MT as '*attracting business through marketing*'. It was '*travel from home to another country for treatment where the host country derives financial benefit from their travels*'. The difference was in their attitude towards competition and benefits.

The corporate sector physicians emphasised three main reasons for the popularity of MT: '*providing opportunities to overseas patients to avail the hi-tech medical facilities in India at a low cost, through skilled physicians who can compete with physicians in the USA and the UK*'; '*quality services are cheaper in India*' with '*culturally similar conditions for neighbouring countries*'; and '*due to post-11 September visa restrictions to the USA, the Arabs and Africans prefer India*'. Some also opined that '*aspects like arrival, pick up, transport, and reception at airports ensure patient satisfaction and help standardise price structure*'. Others went to the extent of proclaiming: '*We deliver caring services*'; '*India is known for its hospitality and there are no racial differences here*'; '*Indians are by nature compassionate and brainy, so we can tap this potential*'; '*We are proud of our medical expertise*'; and '*We have state of the art equipment comparable with the best in the world*'.

Public sector physicians were more detached with a "not-for-us" attitude with lesser involvement in the competition around MT. While they all felt, '*Overseas patients come to public hospital for simpler treatment due to lack of good quality tertiary care in their own country or long queues*', only those in one Delhi hospital with greater exposure to foreign clients felt, '*Medical tourism is a new term for an old phenomenon as our hospital has been treating foreign patients for long*'; '*The volume of overseas patients is not high and they pay like the private patient*'. Use of the term tourism with clinical work made these physicians uncomfortable. '*Tourism is always pleasure related, whereas MT is not*'; '*only traditional healing centres could be identified for promoting health tourism*'. Thus, the majority underplayed the commercial underpinning of the word 'tourism', where the notion of profit from service was critical as they saw them as reinvestments!

Table 1. Distribution of physicians interviewed across type of tertiary care hospitals in three cities

Cities	Delhi		Hyderabad		Chennai	
Physicians	Public	Corporate	Public	Corporate	Public	Corporate
MT Centred Specialities*	16	12	7	6	-	19
Specialities with or without MT**	9	1	2	2	-	7
Administration	3	2	2	3	-	
Total	28	15	11	11	-	26

*Specialities from Cardiology, Orthopaedics, Surgery and Paediatrics.

**Specialists from General Medicine, urology, neurology, gastroenterology, oncology Geriatrics and Preventive and Social Medicine.

3.2 Professional perceptions on work load and conditions in tertiary public and corporate care hospitals

Long queues for OPDs in public hospitals without a place even to stand, beds occupied by more than one patient in three out of four hospitals, stood in contrast to the decorated spaces and low patient load of the corporate hospitals. Public hospital OPDs had 1,500- 8,000 patients every day as compared to 100 to 1,200 per day in corporate hospitals. The bed strength of the former ranged between 1,500- 2,500, while the beds in corporate hospitals varied from 130 to 300.

In public hospitals, inadequacy of funds, political interference in recruitment and promotion leading to mistrust, and undermining the autonomy of institutions were a constant refrain. The internal problems reported were: unhealthy work culture, inefficiency of staff, trade unions without a strong sense of responsibility. A self-critical reputed senior doctor said, '*Physicians lack spirit of academic pursuits, transparency and autonomy in research in a disturbed work culture, therefore, quality of care gets affected and outputs are often irrelevant to our health needs. There is also a disjunction between what we do and what our national priorities are*'. Others agreed that these problems affected medical care for the poor negatively. '*Patients from other states, especially from rural areas,live on roads or Dharamshalas* (*inns*), *paying Rs20-Rs100 per day. CGHS patients wait for surgical treatment despite doctor's orders for admission, for lack of beds. The wait is on an average 6 months - sometimes patients die. There is paucity of blood and Social Welfare Fund schemes are often misused by administration and politicians as few people know of these*'.

Dissatisfaction was expressed by public sector physicians in many ways, '*We don't have adequate medical facilities for the poor in India; without subsidy or insurance 70% of the population cannot afford super specialty treatment*. '*Physicians are mostly overburdened*', '*We have to first meet proper standards of bed and nurse availability*'. '*The working conditions are so poor that, at the end of the day, it is the love for one's work which sustains one in such an environment and individuals focus only on what they can do, forgetting the macro picture*'. '*Public sector has reached a point where a deliberate policy to improve it is unavoidable if standards have to rise*'. A self critical perspective was, '*the patient has no rights, at times we are arrogant, so do not seek active participation from the patient in choosing from the options available to him*'.

In contrast, the corporate sector physicians were satisfied with their working conditions. They emphasised the need to work in a technically equipped work place where they get job satisfaction and are able to practice scientific medicine. They pointed out that their institutions offered differential packages for different income groups. Another advantage they saw was in using their skill to the maximum by taking up consultancies in more than one hospital.

3.3 Public sector doctor's perception of Mt

Out of the 39 interviews of public sector physicians, three sets of opinions emerged on the promotion of MT in public hospitals discussed below. The majority, i.e. 29, were very positive about the scope of MT in India. 14 of these (35.8%) thought it should be promoted even in the public sector while 15 (38.4%) felt public sector was in no position to compete and only the private sector could take advantage of it. Only eight physicians (20.5%) were critical of MT as a strategy. Two (5.1%) did not comment.

3.4 Against MT in principle

The eight physicians against MT in principle, considered MT as, *'a crime against the ordinary Indians,an undemocratic policy', 'practiced without any consensus or meeting of all parties'. 'There is a need to look at our own health needs against which the idea of MT in public hospitals – that were not meant for making profits – makes no sense.' 'Have we treated all Indians that we want to invite patients from abroad?' 'The poor do not get any benefits from the Foreign Exchange.' 'The promotion of MT will not improve the country's health when we are not able to provide basic services to the people'*. They added, *'Our hospital focuses on education, research, and services and will resist money making policies like MT'*. They believed that, *'Medicine is for services and not for business. MT is against the oath of Hippocrates and therefore unethical'; 'it undermines the 'very basis of faith between physicians and patients', 'our institution has the capacity and quality to promote MT, but the institution is not meant for private service... If we spend so much time and energy on few patients it will deviate from the norm/protocol of the institution'*. Proud of the public sector's ability to provide hi-tech 24 hour care without profit, they said, *'Nobody has the time to provide ancillary services, when key services have to be ensured*!' Facilities like deluxe rooms and personal security, required by medical tourists, were inconceivable for them.

'Resource mobilization by the State', 'political will, appropriate priorities, regionalisation and internal tourism' was their answer to inequity, reflecting an acceptance of some degree of the commercialisation which they criticised otherwise. Despite their frustration with state policies, and their view that, *'MT takes advantage of lack of proper regulatory authority or effective laws and implementing agencies and makes malpractice possible, both in the use of technology and pricing'*, they barely contested the State support to MT in the private sector, or pointed towards its international business connections.

They wanted cross border medical treatment for humanitarian reasons and to promote international relations and not MT. For them the issue was not competing for patients in the medical market, but their inability to fully provide for their own patients and the patients' inability to afford payments. *'The Ministry and the government want us to generate our own revenue but we are not getting into the market'*. This moralistic admonition of commercialisation was reflected by a doctor's outburst, *'Our ethics are bad, we let Indian physicians go abroad and invite patients from abroad when we ourselves have surplus patients'*.

Thus 35.8 percent of the public hospital physicians felt they could use MT to the benefit of patients and professional while 38.4 percent felt that in the present form public hospitals were ill suited for MT and would further deprive the general patients, 20.5 percent were in principle against MT in public hospitals.

3.5 MT in corporate hospitals alone

Fifteen public hospital physicians considered MT a legitimate function of the corporate hospitals (with their 'spare capacity'). Public hospitals were seen as *'ill equipped'*, *'already over burdened'*, and *'ill suited'* for it. The objections were two. Firstly: *'MT is going to create a divide by diverting resources from the poor who will suffer more, it will enhance the burden of the poor patients unless the government separates the two sets of institutions'*; *'The onus of care for the poor lies with the public sector and it should look after the interests of the general public'*. Secondly, they argued that: *'Given the present level of investments in public hospitals it is out of the question to achieve the standards needed for MT'*; *'The importance given to medical tourism will certainly affect general health care and public hospital is for all and not for some specific people'*. A doctor said, most physicians expressed their anguish at the state of Public health services, and emphasised the inadequacies, which need to be catered to before embarking on MT. They were clear that, *'MT can be encouraged and promoted in corporate and private tertiary hospitals and it would help the economy and the growth of hi-tech in India'*.

3.6 MT both in public and corporate institutions

Nine specialists argued that India has the best low cost medical expertise in South Asia for treating the overseas patients and generating high revenues. They saw a tremendous scope for MT especially in plastic surgery, orthopaedics, and other non-invasive techniques. *'Wherever these departments are well established and of international standards, they can be opened for MT.'* In Hyderabad, one of the public hospitals was planning to open a 7th floor for foreign patients to promote MT - especially for NRIs. The remaining five specialists felt that primary and secondary health care should be under the public sector, while tertiary care should be through private sector and the pay clinics of government medical colleges and teaching institutions which should provide subsidised/free care for the poor.

The inter-state movement of patients from less developed states and towns to the metropolitan cities was viewed as yet another possibility for internal MT and an opportunity to improve hospital infrastructure, services, and research facilities within the public sector tertiary institutions. Some also argued that, *'A government doctor works from 9 a.m. to 4 p.m. and rest of the time many physicians do not have anything to do but are not allowed to practice privately. Why does the government not utilise the idle hours of the physicians who want to work extra time by opening 'pay clinics' in the evenings? Pay clinics can be operated in the same hospital by developing good infrastructure. In this way, the government will get substantial amount of revenue. This can be utilised for development of existing departments. Otherwise, the government is losing a good hand'*. The contention was that, *'In the government hospitals, there is a need to have more incentives'*. Others disagreed and argued that, *'Physicians in public hospitals work approximately 12 hours a day and to promote MT we need to invest in new infrastructure and manpower as the same set of people cannot meet the additional load'*. This was critical so that, *'we do not bleed existing resources, both human and mechanical'*. It was also claimed that, *'This would not be because of any lack of professionalism in the public sector but due to their poor infrastructure and support systems. The public hospital environment is too dirty to attract foreigners, in fact, it prevents people from coming'*.

They also wanted to have hi-tech facilities like telemedicine which they considered necessary - *'if patients have to be managed at a distance'*, *'be able to compete with corporate sector physicians'*, and *'see different types of patients'*. Further, *'quality of care may improve because of the need to meet western standards'*. This set of experts wanted to simplify laws of organ transplant and make donors easily accessible. They felt that people come to public sector hospitals because of the doctor's name; for MT, the entire infrastructure has to improve along with *'quality of care, values and ethics'*.

They argued that the public hospitals too can generate revenue through MT by overcoming some infrastructural deficiencies. Holistic healing and preventive care should be given priority under MT and they suggested package deals of *'services with free accommodation'* to Medical Tourists. The notion of *'generating money'* for the public hospitals through MT was generally accepted.

Interestingly, those who supported adoption of MT and welcomed markets in public sector medical care did accept that the public sector hospital is the only recourse for the poor. They argued that improving infrastructure will help the poor as well, since the profits would be ploughed back into their care if it could be *'ensured that no discrimination happens between patients'*. The main arguments for promotion of MT in public hospitals were that: a) it would lead to healthy competition and inter-state tourism; b) equipment and infrastructure would be updated and new technology could be brought in for all; c) a learning opportunity for physicians; d) improved quality of care for all; f) revenue generation for the government; g) arrest brain drain from public hospitals; h) a cure for poor administration and inadequate working conditions.

3.7 Perception of physicians in corporate hospitals

90.4% (47) physicians of the corporate hospitals saw a huge potential for MT. None talked about the public sector, which they said, *'is the responsibility of the State and is not our concern'*. A minority of 9.6% i.e. five physicians were critical of MT and were upset with its promotion, they reflected, *'given the poor medical facilities in many parts of the country, the first important thing to do is to indulge in some soul searching before inviting people from abroad'*; *'we need to serve our own people'*, or that, *'MT is like selling India in its naked poverty'*. One specialist said, *'International insurance companies will really benefit from the idea by saving costs and time of the insurance cover, they will look for cheaper options'*. Two of these experts, in a missionary tertiary care hospital, added, *'we are trying for ISC* (*International Standards Certification*) *to promote MT... for private hospitals what matters is money*!' Their views were however overshadowed by the MT enthusiasts who felt, *'An excellent concept that can put India at the top by focusing on specialty treatment'*. The range of concerns expressed by them was the following:

3.8 Finding clients

'If Thailand can attract 2 million medical tourists per year, India should attract 20 million', said the Chairman of a leading corporate hospital. Another vision was to see India as, *'the health care hub of the east'*. The logic, *'India has human capital, is human friendly, and has an infrastructure*

which can be expanded. India needs to progressively exploit this potential'. Another opined, '*MT has picked up in Mumbai, and Delhi is not far behind*', and that, '*it would be beneficial for the country*'. An invitee to a consultation on Britain's National Health Service by the NHS, UK, he had recommended that NHS should send patients to India for cheaper treatment. He said, '*It will be the underprivileged population in UK that would be sent to India for medical treatment. It's a win-win situation forprivate hospitals to attract patients, earn money, be competitive, enhance their standards of care, and generate employment*'. In the process the bigger medical corporations might wipe out the smaller less competitive institutions he added. However, estimating the vast potential of MT, another doctor opined '*In Britain alone, there are 10 million patients who are in the waiting list in NHS, out of which only 11% are British and rest are ethnic people with 80% having private insurance. This is the group which we need to cover - ethnic and privately insured*'. Others proposed targeting the uninsured overseas clients and holding health festivals in client countries like Oman and Maldives.

3.9 Benefits of MT

These experts saw innumerable benefits of promoting MT, specially, in areas such as dentistry, cardiology, orthopaedics, cosmetic surgery, and paediatrics. The most obvious benefit was, '*earning foreign exchange*' or '*more the patients, more the income, which will eventually lead to growth of business.*' Many felt, '*With up- gradation of facilities, health service standards will go up automatically*'. MT was also '*an opportunity to treat different sets of patients having different clinical problems to acquire knowledge and hone our skills*'. It was '*a boon to learn new techniques and handle more complicated and as yet incurable diseases*', with the incoming technologies. They argued that,' *60% of employment is generated through service sector all over the world, so, by making our institutions globally competitive, we can generate employment and retain competent physicians*'. MT, they said, would also create a '*friendly international atmosphere*': its commercial value required that its growth be '*exponential*'!

Other ways suggested to promote MT were: by standardizing and getting accreditation, direct flights to the cities, improving roads, easy money transfers, medical visas and systematic and planned scaling by the state as is being practiced already.

3.10 Physicians view of implications of Mt for medical care in India

Most experts in corporate institutions saw their own sector as the most critical in medical care and earnestly believed that, *'by selling Indian health industry abroad, and by marketing health care, one can increase the earnings and utilise that money in improving the existing infrastructure'*. *'This would force the inefficient public sector to compete and thereby raise its standards and improve job opportunities'*. This set obviously had no consideration about where the public sector resources and their own subsidies came from. Only a minority of five accepted that, *'by getting patients from abroad, dollars will get priority and the less privileged Indian patients will take a backseat'* but they did not talk of implications for public sector.

Among those for MT, the perceived implications were positive as discussed earlier. Some even argued, *'MT may actually be stopping the brain drain of physicians going to work abroad by opening avenues in private sector'*. Others felt, *'Infrastructural improvement will improve services for all'*.

Experts against introduction of MT in public sector pointed out the negative implications of MT, such as: reduced medical care for the poor due to diversion of manpower to the care of overseas patients within the institution, as well as *'probable rise in malpractice by corporate hospitals by not fulfilling the condition of 25% free admissions to patients and thereby increasing the load on public hospitals'*. Thirdly, in MT, *'the preventive component of medicine is left behind'*.

Yet another concern was brain drain out of the public hospitals. Even though it was said, *'We cannot stop them as there is a difference in salaries'* or, *'As such there is no mass exodus'*, Others reaffirmed *'an internal brain drain'*; *'a lot of our expert staff, physicians and paramedics leave public hospitals for the private sector'*. *'Most of the physicians working in XXXX,[a] XXXX are from government hospitals'*. *'Our hospital gets some of the best junior residents from all over the country, and it is used like a nursery for training before they join the private sector'*. Other than money, the reason for leaving the public sector was said to be *'eagerness to learn more and use sophisticated technology like robotic surgery and not getting stagnated'*. These views matched the finding of the study as more than fifty percent of physicians working in corporate hospitals had earlier served in public hospitals. In Chennai, 13 out of 23 physicians (3 did not respond); in Hyderabad, 5 out of 11; and in Delhi, 10 out of 15 had served in public hospitals.

Among opponents of MT there were some who did not think it would have any impact as their institution was capable of fairly handling its patients. Others said, '*If it is only in private hospitals then, that being a separate sector, there is no implication for general patients*'. According to them, promotion of MT in private sector may, in fact, help the poor as the growth of private sector diverts the middle and upper middle class patients away from public hospitals, which can then exclusively cater to the poor. '*Rather than the poor, the rich will be affected by MT, as many times, preference is given to foreign patients who have higher paying capacity rather than rich Indian patients*'.

These views were no doubt contradictory among the physicians for and against MT across the two sectors. Even among the opponents of MT in the public sector, the articulation of the significance of shifts of state subsidies and policy of commercialisation was muted. They saw their challenge as one of ethical practice, getting State resources and problems of micro-management.

4. DISCUSSION

Historically, the middle class – especially physicians and bureaucrats, tend to play a key role in the evolution of policy for medical care [11]. Our study reveals certain trends in the perceptions of selected senior medical physicians that are critical: as pointed out earlier this group of physicians contributes to the formation of public opinion, national policy and programs and shapes institutional traditions and organisation.

4.1 Doctor's Perceptions: differences and similarities & ethical underpinnings

An overwhelming majority (90%) of physicians in the private sector held the State responsible for the care of the larger population. In contrast, the public sector physicians' daily challenge of saving lives within the constraints of their patients' economic status and the meagre facilities available, made them sensitive to the problems of the common man. The glamour of hi-tech associated with MT however dazzled most physicians of both sectors. All physicians of the public hospitals under study felt responsible for the poor, 59 percent (23) acknowledged the irrelevance of MT for common patients and potential problems with MT in public tertiary hospitals, while 36 percent (14) thought it would improve overall patient care due to added revenues 5.1 percent (2) remained non-commital.

The economic and political linkages and dynamics between public and private sectors were largely missing from the doctor's perceptions which was primarily limited to micro management of institutions. The disconnect between their concern for the common man and their views on MT was due to the lack of appreciation of the continuum between commercialisation and MT, and the link between denial of resources to public hospitals and shift of subsidies to the private sector [12]. The profit motive behind MT was recognised, hence, those who did not favour MT did it mostly on moral and ethical grounds.

Interestingly, only the private sector physicians, and those public sector physicians who supported MT in their institutions, talked frankly about the existing and possible links with the international market and its potentials for earning revenues, the need to consolidate the medical industry and of State subsidies to push MT without acknowledging the negative impact of the shift of state subsidies and of the logic of market forces which undermines epidemiological priorities. In contrast those in the public sector opposing MT in principle, underlined the importance of policy changes to mobilise resources for the public sector and rebuild the culture of service.

Private sector physicians were very articulate about their views on the role of the State. They demanded State support to promote MT and believed that the "competition" thus offered would force the public sector to improve! This contorted logic completely ignores the fact that the public sector was set up with State subsidies to be able to run services that are non-competitive and based on epidemiological needs, and that costs were standardised by the State to achieve equity in access. Without State subsidies it would collapse, as commercial competition shifts the emphasis to services like Obstetrics, Orthopaedics, Cardio-thoracic surgery transplants, and Urology etc. and their required manpower that distort priorities within public hospitals and health system as such [8, 13,14]. The public sector experts, though wary of perpetual shortages, accepted the State's indifference and had little to say about its constraints though, a few of them talked of reforming the medical care system.

Unaware of the massive indirect shifts of resources that are being proposed [12], the physicians in this study saw public and private sectors as discrete entities. Almost all (94.7%) opposing MT in public institutions granted corporate sector, the right to use MT to generate revenues as they believed, *'It adds to India's economic growth'*. They did not ask how inclusive is this strategy for economic growth except for one who believed, *MT operates outside the public system, the Foreign Exchange earned stays in the private sector, and the country's poor do not ever benefit*'. The others did not raise the issue of declining resources due to state support to private sector. These views illustrate that the physicians were not familiar with the shifting overall pattern of state financing of health services and it's links with private sector, so well known to analysts of the health service system [7, 15]. Most of them in one voice spoke of the huge potential of MT in India as a vehicle for economic growth while only 10 percent in the private and 20.5 percent in the public sector expressed discomfort with the very strategy of MT in their institutions.

Seen in the framework of domains of micro ethics (relationship with, care, autonomy and choice of patient) and macro ethics (issues of equity, justice, resource allocation, care of the vulnerable and maximum benefit for largest number) provided by Nancy M. Baum [16] and Calman and Downie [17], our findings indicate that private tertiary sector doctor's concerns are largely limited to the micro ethical domain related to standards of care and management, regulations, doctor patient relations, beneficence etc. Only 10 percent of them referred to some macro ethical issues of MT as part of public health-such as needs of the majority and concern for the poor. It was primarily the public sector physicians who at all mentioned macro-ethical issues covering financial constraints, equity for the most vulnerable sections, priorities in technological choices for provisioning of services in public institutions of tertiary care and need for primary health care. Even among them 38.4 percent who wanted the state of the art technology and MT to mobilise revenues for the general patients- like 90 percent of their counterparts in the private sector had no problem with medical care becoming commodity and a part of the market. This however was under the impression that MT can mobilise resources for general patients. Thus, they saw no conflict in macro ethical issues and MT.

4.2 Implications for state policy

A weak professional resistance to MT among senior physicians gives the state a better chance to push forward its neoliberal policies and MT, a logical outcome of the market principle, now extends the boundaries of tertiary health care market. The projected Rs.8,50,000 crores contribution to the GDP by 2020, if the Government supported it, prompted the Union budget for 2003-2004 to give infrastructure status to tourism, paving the way for long term State subsidies [18]. MT is thus a beneficiary of this support while the allocations for health have remained below 1.2% of the GDP over the National Five Year Plan periods.

The State itself promoted establishment of corporations of corporate hospitals to manage MT, as in Maharashtra and Gujarat. The Gujarat government announced its Medical Tourism Policy in December 2006, linking its profits with those of the insurance system and professing an annual increase of 33% in MT growth. To promote this 'business' it proposed a focus on medical education, manpower development (as of physiotherapists, optometrists, pharmacists and nurses), state of the art diagnostics, blood banks, alternate systems and medi-cities. It set up a high level business council under the Chief Minister to plan the promotion of MT [19]. Thus, MT also became an official reason for supporting medi-cities as in Gurgaon (State of Haryana), Kolkata, and Bangalore and professional attraction towards hi-tech helped the state in this expansion.

4.3 Evidence negating perception of advantages of Mt

The corporate hospital consultants saw increase in GDP, business, employment, professional advancement, competitive spirit, international standard of care with advanced technology, and improved infrastructure as advantages. All of this matched the State's vision, and the satisfaction of the Indian elite with economic growth rates and its world class medical care is seen as evidence of success. However, debates on the destructive impact of the corporatisation of medicine, the emergence of MT, and social responsibility/irresponsibility of the corporate sector [20] show the other side of reality. But an understanding of the roots of MT and its larger implications evaded the majority of our respondents. Several serious implications have been recognised in the literature:

- Direct and indirect shift of subsidies for corporate sector. The former itself is calculated to be Rs.57,000 crores [14].

- MT is overshadowing the poor state of Indian public hospitals as the 11th Plan focused on the former and glamorised the institutions catering to MT [3].

- Rise in prices of medical services is making it even less accessible and more inequitable. The logic of price rise is the need for profit maximisation to break even the initial investment of around Rs.100 crores for a 200 bed corporate hospital. It takes some 4-5 years to break even and some 7-8 years to make reasonable profits. Investment in hospitals is characterised by low returns – the profit rates are around 13% lower than that in IT, finance, or retail [21].

- Rising numbers of establishments of profitable super specialisation and focus on life style surgeries, assisted reproduction, orthopaedic, eyes and cardiac services are distorting priorities as epidemiological reality becomes irrelevant for investors [22].

- Overemphasis on tertiary care is leading to shortage of skilled work force in primary care and distorted manpower planning [23].

- Brain drain both external and internal [24].

- Hi-tech institutions act as a push factor for insurance businesses which keep the poor out in India yet, the potential for earning revenues through MT will become an important argument for private hospitals demanding more subsidies from the government in the long run and further undermining services for the poor [25]. Yet there is evidence to show lack of transparency and unethical practices in these hi-tech institutions in India [6].

These facts however, did not constitute the basis of perceptions of even the 27.5 percent of those who were in principle against MT in public sector tertiary hospitals or the 9.6 percent of the private sector physicians who felt it did not help the poor. Both rejected MT on ethical and experiential grounds, the former more openly than the latter.

Despite the fact that patients from the developed world seek MT to save costs in elective treatment (plastic surgery, cosmetic, dental and wellness treatment) or if they are under and uninsured, MT may not be a solution forever. If the efforts of developed nations to restrict losses they incur by cross-border patient travel, by creating their own security systems, succeed, then the profitability of MT may decline. Developed nations are changing insurance laws for those who use MT [10], and developing collective mechanisms of cross border treatment at regulated prices as in the European Union [26] or introducing health Reforms as in the US [27] to handle the crisis of health services. What will then happen to the huge five star high-tech hospitals in India, once their clientele falls? Already many of them survive on heavy State subsidy and by converting hospitals into restaurants, pastry, gift and glossy book shops. Even when it contributes to the growth-oriented economy, its benefits exclude the majority. Is it then an impending disaster, the costs of which will be shifted on to the State and the exchequer as in the case of the Commonwealth Games [28]?

Under these rather alarming circumstances, where India's health service system is responding to the needs of an elite international and national community, the role of physicians committed to the majority of Indian patients acquires great significance. Their silence and indifference can harm the common patients. Therefore, an over-arching understanding of MT and its implications for the public health system is crucial. It helps one to realise that MT is for cross-border trade in medical care, transfer of technology, manpower, and knowledge that favours the elite and not necessarily the ordinary users of public sector tertiary hospitals. This realisation then provides a more objective basis to their demand for essential policy changes.

ENDNOTE

[a]All identifying information has been removed and replace with XXXX to ensure anonymity.

ACKNOWLEDGEMENTS

We are grateful to the School of Social Sciences, JNU and MHSP project (Monitoring shifts in health sector policies in South Asia) for the financial assistance to make this research possible. We would also like to thank researchers Francis Adaikalam, Sutanya Mahapatra, Srilaxmi Divakar and Ashima Jena for helping us in data collection.

AUTHORS' CONTRIBUTIONS

Both the authors IQ and SR have contributed equally to the study, from conceptualising the problem, review of literature and designing research and formulating the tools. Under the guidance of IQ, SR along with the research team conducted the field work. Both the authors have analysed and written the paper. Both the authors have read and approved the final manuscript.

REFERENCES

1. CII-McKinsey: Health Care in India: The Road Ahead. CII. 2002, New Delhi: McKinsey and Company and Indian Healthcare Federation

2. Planning Commission:: Tenth Five Year Plan, 2002-2007. 2002, New Delhi: Chapter on Health, Government of India, Volume 2

3. Planning Commission:: Eleventh Five Year Plan, 2007-2012. 2008, New Delhi: Chapter on Health, Government of India, Volume 2

4. Government of India: Ministry of health and family welfare, draft national health blll. 2009

5. Planning Commission:: Twelfth Five Year Plan, 2012-2017. 2013, New Delhi: Chapter on Health, Government of India, Volume 3

6. Qadeer I, Reddy S: Medical care in the shadow of public private partnership. Social Scientist. 2006, 34 (9–10): 4-20.

7. Baru RV: Privatisation of health services: a south asian perspective. Econ Pol Wkly. 2003, 38 (42): 4433-4437.

8. Chanda R: Trade in Health Services. 2010, http://www.who.int/trade/en/THpart1chap3.pdf webcite,

9. Budget 2003- 2004: speech of Jaswant Singh, minister of finance and company affairs, Feb. 28 2003. 2013,http://indiabudget.nic.in/bspeech/bs200304.pdf Webcite

10. CII: Great Advantage in India Medical Tourism: CII August 2004. 2013, http://www.medicaltourindia.com/medical-tourism-news.asp Webcite. Accessed on 5th May 2013

11. Banerji D: Landmarks in the development of health services in India. Public Health and the Poverty of Reforms. Edited by: Qadeer I, Sen K, Nayar KR. 2001, New Delhi: Sage Publication

12. Reddy S, Qadeer I: Medical tourism in India: progress or predicament?. Econ Pol Wkly. 2010, 45 (20): 69-75.

13. Whittaker A: Pleasure and pain: medical travel in Asia. Global Public Health. 2008, 3 (3): 271-290.

14. Chanda R: Trade in health services. Bull World Health Organisation. 2002, 80 (2): 158-163.

15. Sengupta A, Nundy S: The private health sector in India. BMJ. 2005, 331: 1157-1158.

16. Baum NM, Sarah EG, Susan DG, Peter DJ: Ethical issues facing public health practitioners. American Society for Bioethics and the Humanities Annual Meeting: Connecting and Collaborating 10 (19). 2007, 7-Washington, DC

17. Calman KC, Downie RS: Ethical principles and ethical issues in public health. Oxford Text Book of Public Health. Edited by: Roger D, Robert B, Mary AL, Martin G. 2011, London: Oxford University Press

18. Medical tourism policy. 2010, http://www.indiacore.com/tourism.html webcite,

19. Gujarat medical tourism policy: draft. 2010, http://www.discoveredindia.com /gujarat/tourism -sectors-in-gujarat/medical-tourism-in-gujarat.htm webcite,

20. Sengupta A: Medical tourism in India: winners and losers. Indian J Med Ethics. 2008, 5 (1): 4-5.

21. Kumar P, Simi TB: Barriers to movement of healthcare professionals a case study of India. 2010, http://www.cuts-citee.org/pdf/RREPORT07-03.pdf Webcite,

22. Venkata K, Prasad CB: Medical Tourism Industry - Advantage India. 2010, http://dspace.iimk.ac.in/bitstream/2259/483/1/255-260.pdf Webcite,

23. Hazarika I: Medical tourism: Its potential impact on the health workforce and health systems in India. J Health Plann Policy. 2010, 25 (3): 248-251.

24. Hopper CR: Adding insult to injury: the healthcare brain drain. J Med Ethics. 2008, 34 (9): 684-7.

25. Gupta SA: Medical Tourism in India: winners and loosers. Indian Journal of Medical Ethics. 2008, 14 (1): 4-5.

26. Bertinato L, Busse R, Fahy N, Quigley HL, McKee M, Palm W, Passarani I, Ronfini F: Policy brief cross border health care in Europe. 2010, http://www.euro.who.int/__data/assets/pdf_file/0006/108960/E87922.pdf Webcite,

27. Health care reforms to impact medical tourism. 2010, http://www.imtjonline.com/news/?EntryId82=129713Webcite,

28. Chotani S, Sultana S: Common Wealth Games in Delhi: How much does national prestige cost? A report; 2007. 2007, New Delhi: Monograph of The Hazards Centre

Chapter 10

Knowledge brokers, companions, and navigators: A qualitative examination of informal caregivers' roles in medical tourism

Victoria Casey[1], Valorie A Crooks[1]*, Jeremy Snyder[2] and Leigh Turner[3]

[1]Department of Geography, Simon Fraser University, Burnaby, British Columbia, Canada.

[2]Faculty of Health Sciences, Simon Fraser University, Burnaby, British Columbia, Canada.

[3]Center for Bioethics, School of Public Health, and College of Pharmacy, University of Minnesota, Minneapolis, Minnesota, USA.

ABSTRACT

Introduction

Many studies examining the phenomena of medical tourism have identified health equity issues associated with this global health services practice. However, there is a notable lack of attention in this existing research to the informal care provided by the friends and family members who typically accompany medical tourists abroad. To date, researchers have not examined the care roles filled by informal caregivers travelling with medical tourists. In this article, we fill this gap by examining these informal caregivers and the roles they take on towards supporting medical tourists' health and wellbeing.

Methods

We conducted 21 interviews with International Patient Coordinators (IPCs) working at medical tourism hospitals across ten countries. IPCs work closely with informal caregivers as providers of non-medical personal assistance, and can therefore offer broad insight on caregiver roles. The interviews were coded and analyzed thematically.

Results

Three roles emerged: knowledge broker, companion, and navigator. As knowledge brokers, caregivers facilitate the transfer of information between the medical tourist and formal health care providers as well as other staff members at medical tourism facilities. The companion role involves providing medical tourists with physical and emotional care. Meanwhile, responsibilities associated with handling documents and coordinating often complex journeys are part of the navigation role.

Conclusions

This is the first study to examine informal caregiving roles in medical tourism. Many of the roles identified are similar to those of conventional informal caregivers while others are specific to the transnational context. We conclude that these roles make informal caregivers an integral part of the larger phenomenon of medical tourism. We further contend that examining the roles taken on by a heretofore-unconsidered medical tourism stakeholder group sheds valuable insight into how this industry operates and that such knowledge is necessary in order to respond to the health equity debates that surround this particular global health services practice.

KEYWORDS

Informal caregiving; Medical tourism; Global health services; Healthcare providers; Qualitative analysis; Family caregivers

1. INTRODUCTION

Health care delivery involves many provider groups, both formal and informal, that address different facets of patient care. Doctors and nurses are examples of formal provider groups in that they receive specialized training specific to the care they deliver, are paid to deliver this care, and are commonly licensed professionals [1, 2]. Much social health research has documented the important and complex responsibilities that formal providers assume in ensuring patients' health and general wellbeing e.g. [3, 4, 5]. In contrast, informal health care providers include friends, family, and some volunteers. Members of these groups provide essential care despite the fact that they are not formally trained and paid health professionals [6, 7]. These informal care providers, or caregivers, often provide ongoing essential care in the home, such as administering medications, managing wounds, and assisting with rehabilitation [8]. Meanwhile in hospitals and other types of residential care settings, they perform tasks that augment services provided by on-site formal providers. Examples of such tasks include monitoring symptoms and articulating patients' preferences to health care professionals [9, 10, 11]. The roles that formal and informal providers fill and the responsibilities assigned to each collectively work together to enable health care delivery. In this article, we examine the caregiving roles assumed by a specific group of informal health care providers: the friends and family members that accompany medical tourists abroad for private medical care.

Hospitals, clinics, and individual health care practitioners who provide medical treatment to privately-paying traveling international patients who do not have a formal cross-border care referral are part of what is often described as the 'medical tourism industry' [12, 13, 14]. This industry is reported to be on a multi-billion dollar scale and involves the transnational movement of patients to a growing list of destination countries located on nearly every continent [13, 15, 16, 17, 18]. There are a great variety of procedures that can be obtained in medical tourism facilities, including, though not limited to, cosmetic surgeries, cardiac surgeries, and orthopedic surgeries [15, 19, 20]. Significant concern has been raised that the medical tourism industry is exacerbating health inequities in destination countries through such measures as recruiting formal health care providers out of public systems and into private ones, which prices locals out of needed health care [21]. Supporters of the industry meanwhile suggest that the practice of treating privately-paying international patients can assist with addressing health inequities through measures such as retaining formal providers who may otherwise have left the country in search of higher pay or more technologically sophisticated practice and bringing in capital [21].

Research examining medical tourism has paid some consideration to numerous health care provider groups central to this global health service practice. Perhaps not surprisingly, much of this research examines formal health care providers rather than informal ones, though sometimes members of the latter category are mentioned in passing. Although academic studies and industry reports note that it is common for medical tourists to be accompanied by friends and family members (e.g., [6,16, 22, 23, 24]), they are rarely framed as informal health care providers or as a stakeholder group meriting additional exploration. Meanwhile, it is widely acknowledged that informal caregivers provide the bulk of patient care throughout the life course and across the care continuum and play an important role in establishing care recipients' health outcomes [7, 24, 25, 26]. Given the importance of informal caregivers in health care provision, it is surprising that their contributions to the care of medical tourists have not attracted greater scrutiny. Only non-academic work has addressed the practice of informal caregiving in medical tourism. Two published narratives of medical tourists' journeys written by their informal caregivers reveal that significant caregiving responsibilities were assumed by these individuals [27, 28]. Their detailed narratives document numerous instances of making decisions on behalf of the patient, liaising with formal providers, coordinating appointment scheduling, offering hands-on care, providing emotional and spiritual support, and taking responsibility for managing care-related finances. These narratives, although not scholarly in nature, are nonetheless illuminating because they suggest that informal providers can play a significant role in the practice of medical tourism. In this article, we aim to identify and explicate the informal caregiving roles that friends and family members assume in the course of a medical tourist's journey. By 'roles' we refer to their key social functions in the maintenance of medical tourists' health and wellbeing and the larger transnational care practice of medical tourism.

Our exploration of informal caregivers' roles herein is informed by insights gleaned from interviews with International Patient Coordinators (IPCs) working at medical tourism facilities. IPCs work at destination facilities; their task is to coordinate medical tourists' care. Their responsibilities include arranging ground transportation and local travel, communicating with doctors, scheduling medical appointments, and providing support and guidance for patients and their caregivers. Because of the nature of their jobs, every year they interact with anywhere from tens to hundreds of medical tourists and their informal caregivers. Given their function, we believe that by sharing their observations and experiences they are well positioned to identify the informal care roles filled by this caregiver group. In the section that follows we provide an overview of the study design and a description of the 21 IPCs with whom we spoke. We then present the findings of a thematic analysis that identified three roles commonly filled by medical tourists' informal caregivers: knowledge broker, companion, and navigator. We subsequently discuss the findings in light of the existing medical tourism and informal caregiving literatures and offer directions for future research. We conclude by reflecting on the relevance of this analysis for providing new insights that have relevance for the health equity debates that surround the global medical tourism industry.

2. METHODS

This analysis emerges from a large, multi-method study that explores first-hand accounts of medical tourists' informal caregivers and those who have worked closely with them in a professional capacity. Here, we report on the findings of interviews conducted with IPCs about their interactions with and observations of these caregivers. The findings speak to the roles that caregivers from a range of home countries fill as they accompany medical tourists seeking a variety of medical procedures at international health care facilities.

IPC recruitment commenced upon receiving approval from the Research Ethics Board at Simon Fraser University. We sought participants from a diverse range of countries and facilities using several concurrent methods: (1) emailing letters of invitation to hospitals and clinics whose websites mentioned IPCs, IPCs identified in online medical tourism directories, and IPCs who had posted on online forums; (2) snowballing out from initial participants; and (3) disseminating calls for participants through our team's networks and online medical tourism industry forums and magazines. Recruitment materials indicated that interviews could be conducted in

English or French. A later request for a Spanish-language interview was also accommodated.

Interested potential participants who contacted us by e-mail were sent an information sheet that provided additional information concerning the study and described their rights as participants including confidentiality. Before this sheet was sent, participant eligibility was confirmed. Because many potential participants did not use 'IPC' as their official job title, they were required to indicate that: they worked with international patients who obtained procedures at medical tourism hospitals or clinics that offered surgical procedures without third party involvement such as organ transplantation; they were physically present in the facility with the medical tourist; they made care and other arrangements; and they assisted clients in a non-clinical capacity. To capture diversity among the sample, no more than three individuals from a single facility were interviewed. We stopped active recruitment and interviewing when we reached our target sample size of 20 interviews, a point that coincided with when new potential participants were no longer being identified.

Interviews were conducted over telephone or Skype according to the participant's preference. They typically lasted for 45-75 minutes. A semi-structured interview guide was used, enabling the capture of issues central to the study's objectives and topics that were important to the participants. The first author conducted 19 English-language interviews while a knowledgeable collaborator conducted one in Spanish. No French-language interviews were requested.

Verbal consent was obtained before each interview, and the interviews lasted for approximately 45-60 minutes. As shown in Table 1, the interviews covered topics such as: (1) informal caregiver characteristics, (2) interactions between caregivers and medical tourism facility staff, (3) caregivers' roles and responsibilities, and (4) the risks to which caregivers can be exposed while travelling with medical tourists and providing care to them.

Twenty interviews were conducted with 21 IPCs (one interview had two participants) and drew from their experiences working at 16 different medical tourism hospitals or clinics in Bolivia, Costa Rica, Barbados, Mexico, the United States, Croatia, India, Israel, Thailand, and Turkey. Twelve IPCs mainly dealt with North American medical tourists, six mostly serviced Europeans, one primarily saw Australians and another Africans, and the remaining two did not report a particular regional orientation. The procedures provided at the facilities where the participants worked included cosmetic surgery, bariatric surgery, orthopedic surgery, oncology procedures, spinal surgeries, veinoplasty, and cardiac surgery.

Table 1. Selected interview questions

Question	Sub-probes
In your experience, what is the typical relationship between patients and their travel companions?	What are some of the common characteristics of travel companions?
What are some of the reasons that you interact with travel companions?	Before arrival? While abroad? Upon returning home?
What kinds of responsibilities do you commonly see travel companions taking on?	In relation to: communication; symptom monitoring; hands-on care; corresponding with friends and family at home; providing emotional support to the patient; providing spiritual support to the patient; making arrangements; travel and tourism activities; other relevant activities?
Have you ever experienced a situation where a travel companion's health worsened or improved while they were abroad (both in hospital or after discharge)?	How common is this? What could be the cause of their worsened or improved health? Can you think of any problems, stresses, difficulties that travel companions face while the patient is in hospital and after discharge?

All interviews except one were recorded digitally and transcribed verbatim. Technical difficulties prevented the exception from being recorded, and detailed interviewer notes were instead used to document the interview. All transcripts and notes were loaded into NVivo, a qualitative data management program, after which thematic analysis was conducted.

The thematic analysis involved six steps. First, all investigators reviewed the transcripts and notes. Second, emerging themes and outliers were identified during a face-to-face meeting with all investigators. Third, the first and second authors created a preliminary coding scheme that identified overall thematic concepts and their components. This involved collaboratively creating tiers of category headings that were manipulated until the authors were confident that they would entirely capture the study objectives as well as the themes and outliers identified in the meeting. Fourth, the first author coded the data in NVivo, with input on code refinement and interpretation from the second author. Fifth, the first and second authors identified emerging trends and patterns relevant to the themes pursued in the current analysis, namely those pertaining to caregivers' roles. Sixth, a refined interpretation of meaning in the coded data was revealed through a comparison of the trends and patterns with existing knowledge and the study objectives [29]. This comparison was initially done by the first and second authors and then confirmed by the full team. Characteristic of thematic analysis, this analytical process enabled common themes to emerge despite the differences in IPC participants' work environments and work histories [30].

3. RESULTS

All IPCs reported that it was common for medical tourists to bring at least one friend or family member abroad with them unless they were specifically discouraged from traveling with a companion[a]. Family members, especially spouses, were the most common type of informal caregivers present at the facilities where participants worked. Facilities that offered in-patient procedures often had cots or beds available in patients' rooms for these informal caregivers. This arrangement demonstrates the intense physical proximity of this transnational caregiving practice. Figure 1 provides an example of this kind of patient room. In the case of out-patient clinics or where co-habitation in the patient's hospital room was uncomfortable or impossible, friends and family typically stayed at nearby hotels or rented apartments.

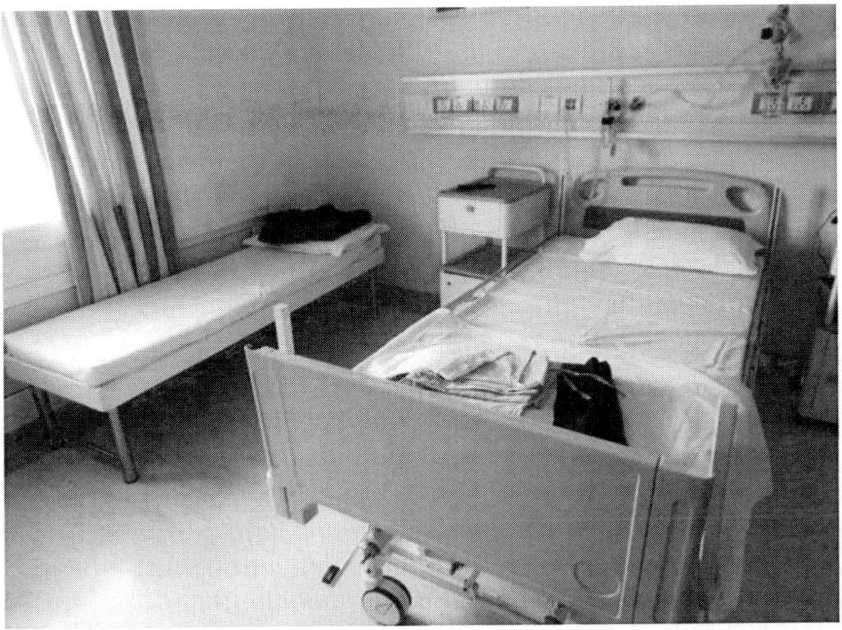

Figure 1. Typical medical tourist patient room. Taken at hospital in India that treats medical tourists, this photo conveys the close physical proximity that is experienced between some medical tourists and their informal caregivers while abroad. The cot on the left is for the friend or family member providing care while the bed on the right is for the patient. (Photo credit: authors).

Thematic analysis of interviews conducted with 21 IPCs about their interactions with and observations of medical tourists' informal caregivers revealed three roles commonly adopted by this group: (1) knowledge broker, (2) companion, and (3) navigator. The knowledge broker role refers to key functions undertaken by caregivers around the transfer of information between the medical tourist and facility staff or others. The companion role refers to functions around the provision of emotional and physical comfort or support as well as hands-on care. Finally, the navigator role encompasses functions dealing with information gathering and care coordination responsibilities assumed by medical tourists' informal caregivers. While IPCs observed that many caregivers adopt all three roles, often simultaneously, we consider them separately in this section in order to provide detailed accounts of the distinct features of each role. We include verbatim quotations throughout in order to 'give voice' to the participants. Each quotation is followed by a country name and a number. The name refers to the country in which the participant worked at the time of the interview while the number indicates the number of years they had worked as an IPC at that particular facility.

3.1 Knowledge broker

Participants emphasized that a key role caregivers play is that of knowledge broker. The facilitation of knowledge transfer from medical tourism facility staff to the patient by caregivers is typically done in four ways: inquiry, clarification, translation, and retention. First, caregivers commonly make inquiries on behalf of patients. Questions directed toward IPCs tend to be about logistics or planning, while those directed at formal providers cover clinical concerns such as prescriptions, care options, and advice for after discharge. Second, by seeking clarification caregivers help to ensure that patients clearly understand medical information. This activity can be especially helpful when patient rights are explained, when complications occur, or when patients are confused about what they have been told. Third, caregivers help with translation or basic communication when they are more proficient than the patient in the language or regional vernacular used by facility staff. "*When the patient comes over here, the language may differ or the accent may differ, [the understanding of] English between the patient and the [facility] staff may differ slightly, so there needs to be someone who can patiently communicate with the [facility] staff [if the patient cannot]*" (India, 2.25). Finally, caregivers retain information that has been conveyed by facility staff to the patients. "*Sometimes when you're in that position [as a patient], you're ill, you're in a bed, the doctor comes in and says x, y or z, you don't remember it... So it's nice to have someone there with you who will be able to retain that information*" (Barbados, 2). Knowledge brokering by facilitating the transfer of information from facility staff to patients can be demonstrative of collaboration between caregivers and formal providers and other facility staff.

Caregivers often engage in knowledge brokering by offering unprompted but useful information about medical tourists to facility staff. IPCs noted that caregivers frequently offer this kind of information while patients are at the facility and also following discharge. For example, caregivers may voice patients' concerns or articulate complaints when patients are reluctant to make such remarks. They may also help formal providers obtain correct information when patients are untruthful (e.g., purposely not reporting their correct weight), do not accurately recall their health history, or are unwilling to communicate. For example, one participant reported that she once worked with a patient who was uncommunicative and therefore the accompanying caregivers answered her questions instead: *"...sometimes the patient didn't feel like talking and then his mother or his wife would give me the information"* (Mexico, 3). These types of information exchanges between facility staff and caregivers, ones that are not prompted by the medical tourist and may not occur in the patient's presence, reveal *"...the good, the bad, the ugly, everything [about the patient that may be helpful to facility staff]"* (Costa Rica, 2.5). In other words, caregivers may broker information or share details with facility staff to ensure that they have accurate information with which to make decisions that affect patients' health independent of requests to do so from medical tourists.

Participants observed that caregivers often act as information liaisons between patients and friends and family members back home. They provide friends and family members with updates by phone, email, Skype, and even letters. One participant recounted that caregivers commonly *"bring their computer [or] they can borrow our telephones and right after the surgery they...communicate with people from home just to let them know that everything went okay"* (Mexico, 3). Acknowledging the importance of this aspect of caregivers' roles, some of facilities even offer them free international calling. This dimension of knowledge brokering serves to underscore the truly transnational nature of the roles that caregivers take towards medical tourists as well as within the larger practice of medical tourism.

3.2 Companion

One way that informal caregivers engage in the companion role is by creating a feeling of emotional safety and security for the medical tourist. A participant reported that caregivers provide *"... emotional support, which was probably the most crucial [thing that they can do]. It's something that a medical tourism company cannot offer"* (Turkey, 1.5). Caregivers are familiar to patients, they have established relationships with one another, and bonds of trust already exist before their journeys begin. These traits put them in a unique position to offer better quality support, including moral and spiritual, relative to facility staff. Caregivers' trusting relationships with medical tourists often makes them aware of patients' preferences and needs. All the IPCs with whom we spoke indicated that trusting relationships between patient and caregiver can have emotional benefits for the medical tourist and that this social bond is a significant reason why they encourage medical tourists to avoid travelling alone. To a medical tourist, a caregiver is "*someone there who would be able to share your...experience...you wouldn't have to...be looking for words, you would just be able...to flow with this person because this is someone that you know, it's someone you have a history with, so you'd be more comfortable*" (Barbados, 2). According to the participants, the presence of the caregiver can be beneficial to creating a feeling of safety and security for the patient.

Many participants noted that caregivers typically try to address medical tourists' comfort in their role as companions. This assistance comes in the form of: providing the patient with wanted or needed items, monitoring symptoms, and helping the patient deal with the 'foreignness' of the destination country. IPCs commonly observed caregivers obtaining items for medical tourists inside and outside the facility because, participants postulated, patients were more comfortable asking their companions than their IPC. For example, patients who dislike the food served at the facility can send caregivers into the surrounding area on regular food purchasing trips. To ensure physical comfort, nearly all participants noted that caregivers monitor patients' symptoms and alert formal providers of noteworthy changes. A participant explained: "*...we provide them with our extension numbers and they might call us just to say '...my wife is having nausea after the operation, do you think somebody can give her some medicine'. So we do quite often see that situation*" (Thailand, 6). Caregivers may further ensure patients' comfort by being a familiar, reliable figure in an otherwise unfamiliar environment. IPCs explained that being outside one's own country can be inherently stressful. Language differences in particular can be especially difficult for the patient. Patients and caregivers "*really just have each other... I mean the outside world speaks a different language, different culture. They rely a lot on each other*" (Turkey, 1.5).

Many participants reported that caregivers commonly provide hands-on care, especially after the patient is discharged from the facility. This care may take the form of assistance with mobility and everyday tasks and help with following clinical advice. Patients' mobility and their capacity to perform everyday tasks after surgery are typically minimal, and therefore some facilities require that medical tourists bring a caregiver to assist them with daily activities following the operation. Caregivers often help patients with dressing, showering, toileting, and mobility at the facility and after discharge. After discharge "*... patients...might need the help of a companion* *with luggage, with getting in and out of a vehicle, perhaps getting up to their hotel room, go out to dinner, things like that*" (USA, 8). Caregivers also help patients follow clinical advice such as taking medications, sometimes offering hands-on care to ensure this advice is followed. One participant explained that caregivers need to know "*what not to do [when caring for the patient]: do not...bend the knee over this position if he has a knee replacement, or do not give him anything to eat besides what the doctor says if he has a gastric sleeve...*" (Costa Rica, 0.6). In some cases, caregivers continue this aspect of their companionship role after returning to their home country, ensuring appropriate post-operative and follow-up care regimens are followed.

3.3 Navigator

Caregivers fill a navigator role when they guide the patient through various aspects of the medical tourism experience. One such aspect is geographical and cultural navigation. For example, caregivers typically gather tourist information. As mentioned above, they may do this by asking facility staff, especially IPCs, questions about particular destinations in the local area. Some of these questions may pertain to tourist activities that are suitable for the patient after discharge. Caregivers may also seek location-specific information from IPCs, such as "*how to take a taxi and...what areas are safe to go to and what aren't...*" (Mexico, 12). In addition to gaining familiarity with navigating the destination country or city, caregivers typically learn how to navigate the destination facility and transmit this information to the medical tourist. According to one participant, caregivers are "*the ones who read through the instructions [about the facility] and enforce them [with the patient]...*" (Mexico, 1.5, 0.6). Participants also noted that, in their capacities as navigators, all caregivers familiarize themselves with useful locations in the facility such as bathrooms and magazine vendors.

Most participants observed that a large part of a caregiver's navigator role involves coordinating paperwork and gathering required documentation. "*Usually the companion, to relieve the patient that's having the surgery, does all the running around to make it happen*" (Costa Rica, 5). Many IPCs reported that completing paperwork is one of the first tasks that caregivers must undertake after arriving at the facility. This task requires completing forms for the patient and verifying that patient information is accurate. They also "*...tend to want to deal with the finances...they're always very worried about [the patient] being worried about...the balances or the costs* *that their stay incur. They tend to want to shield [the patient] from that"* (Barbados, 2). They typically monitor and complete financial paperwork and may access a bank in person or electronically to exchange currency, to ensure funds are available, and to get money for airfare, treatment, and other expenses. Caregivers also commonly transfer documentation from the facility abroad to patients' regular physicians and vice versa before departure and upon return home. These documents include medical records, letters, prescriptions, and/or test results. IPCs remarked that navigating the coordination of paperwork and documentation is generally done by caregivers to minimize the number of concerns or stressors that medical tourists encounter.

Creating plans and ensuring that travel itineraries are followed are both common responsibilities for caregivers in their roles as navigators. Caregivers sometimes play a large role in preparing for the trip abroad. They communicate with IPCs to organize the trip, sometimes as early as sending the initial inquiry. They continue to fill this role throughout the stay by arranging lodging, tours, and food options. They may further help patients by arranging ground transportation and international flight arrangements. "*...They get the wheelchair for them and, you know rent a chair car to transport the patient to and from the hotel...*" (USA, 8). Caregivers also sometimes create itineraries with the help of IPCs, and typically ensure that these plans are followed by the medical tourists. A key reason for this is because medical tourists typically look to caregivers "*for [information about] the whole journey, you know: 'What time are we going to the doctor? What time is our appointment? What time is the check-up? When do we go next? Tomorrow's a day off, are we going touring? Are we going to the beach? Are we going shopping?'"* (Israel, 3). Caregivers know the estimated length of the patient's stay in the facility and note changes in the recovery schedule after surgery so that they can alter or cancel activities as needed. Even after they are back in their home country, caregivers often help patients create and follow an itinerary of follow-up appointments. The majority of participants noticed that caregivers generally help patients navigate their experience abroad by making and altering plans as needed, and that nearly all of them play an important role keeping patients on schedule.

4. DISCUSSION

The analysis has found that medical tourists' informal caregivers adopt three major roles that complement those of formal health care providers: knowledge broker, companion, and navigator. As knowledge brokers, they facilitate the transfer of information between the medical tourist and the formal health care providers, the IPC, friends and family members back home, and others. As companions they support the medical tourist emotionally by ensuring their comfort and providing hands-on care. Finally, in their role as navigators they gain familiarity with the destination country, coordinate the trip, and handle facility documents. In this section, we examine interconnections between these roles and consider all three in light of the existing medical tourism and informal caregiving literatures. We also identify future directions for research emerging from the findings.

Each of the three caregiver roles identified by IPCs encompass several responsibilities. Many responsibilities are unique to a single role. Handling finances, for example, which is part of the navigation role, has no overlap with other role components, though the outcomes of this responsibility might have implications for caregivers' other roles. Other responsibilities might seem less clearly delineated with regard to the roles to which they contribute because of the similarities of the activities they involve. For example, there are information exchange aspects to all the roles. The difference is the intermediary or primary position of the caregiver in the exchange and the nature of the information. Offering or retaining information on behalf of the patient is characteristic of the knowledge broker role. Meanwhile, flagging the patient's changing health status to staff relates most closely to the companion role, whereas discussion about travel logistics relates to the navigator role. By both focusing on the intent of an action being undertaken and having identified the scope of each role it was clear in the analytic process as to which responsibilities were attributed to what roles.

A strong relationship exists between the caregiver roles because the actions, activities, and overall responsibilities undertaken in one role can have implications for the other two roles. In this way, there is overlap between all the roles identified in the findings. By helping patients follow clinical advice, which pertains to the companion role, caregivers act on advice they might have retained in their capacity as a knowledge broker. When the clinical advice requires arranging medical appointments and ensuring that patients arrive at appointments in a punctual manner, then there is also overlap with the navigator role. Knowledge brokering and the companion role further overlap when caregivers monitor medical tourists' symptoms and communicate those observations to a health care provider; symptom monitoring is identified as part of the companion role, whereas voicing comments or concerns about symptoms to formal providers is part of the knowledge broker role. These are but a few of the many examples of the ways that distinctive roles become interconnected through the practice of informal caregiving in medical tourism. Although these interconnections can create some overlaps or redundancies between roles, we believe that the distinctions between the knowledge broker, companion, and navigator roles remains useful for clearly positioning the friends and family who accompany medical tourists abroad as informal caregivers, and ultimately unpaid health care providers, within the industry.

Whereas the existing medical tourism literature does not discuss informal caregivers' roles in detail, there are some mentions of the same activities or responsibilities discussed by the IPCs with whom we spoke. Solomon [31] peripherally mentions caregivers' research responsibilities associated with the initial inquiry as well as the fact that they have in-facility information seeking interactions with IPCs. These findings from Solomon's ethnographic study, though not a central part of his analysis, parallel some of our own research findings. Kangas' [32] ethnographic research into the travel of Yemeni patients to other countries for private medical care confirms the centrality of family members in making the initial decision to access care abroad and their knowledge brokering roles. Kingsbury et al. [33] state that caregivers sometimes need to assume essential decision-making responsibilities and that unanticipated changes in medical tourists' health status can force them to "navigate shifting boundaries of...[their] roles." Two autobiographical narratives written by caregivers about their journeys abroad with medical tourists offer the strongest source of confirmation of the findings reported herein [27, 28]. The authors of both narratives disclosed their active participation in all three of the roles identified here. Indeed, these former informal caregivers engaged in most of the responsibilities attributed to each of the knowledge broker, companion, and navigator roles (see [33] for greater analytic discussion of these narratives). With the findings of this article, we assist with putting discussion of caregiver activities and responsibilities from these other sources into context by explicitly considering the roles to which they are contributing.

The existing informal caregiving literature demonstrates that there are indeed commonalities between the roles filled by medical tourists' caregivers and those adopted by other types of informal caregivers. Much of this literature is focused on the practice of informal caregiving in the home (e.g., [6, 8, 10, 25, 34]). One difference between these caregivers and those discussed in this article is the person who serves as a first point of contact and ongoing informational resource. It is the IPC for medical tourists' caregivers, whereas homecare nurses often serve in this capacity for 'conventional' (i.e., local, non-transnational, long-term) informal caregivers in the home [35, 36]. Furthermore, our findings reveal that medical tourists' caregivers sometimes arrange lodging and tours and transfer documents to and from the destination country, whereas these are not common activities for non-transnational caregivers [8, 34, 37]. Differences such as these can be attributed to the specific context of care (i.e., local vs. transnational care and familiar vs. unfamiliar environment). The greatest overlaps between roles undertaken by medical tourists' caregivers and other types of caregivers come in relation to non-context-dependent functions, such as facilitating information transfer and engaging in symptom monitoring. Many studies of informal caregiving in the home have found that informal caregivers in this setting play a pivotal role in transmitting details on changes in patients' health status to formal health care providers, keeping logs of symptom changes, and other similar activities [11, 37, 38, 39]. There are also cases in which the specific activities undertaken by medical tourists' caregivers are different from those of conventional informal caregivers but the overall roles being enacted are identical. For example, finding a wheelchair to aid in a medical tourist's mobility in an airport is likely done with the same intent as arranging for a patient's accessible transit pick-up at home, a common activity undertaken by caregivers in other contexts [34, 37]; in both cases the caregiver functions as a navigator.

5. LIMITATIONS

This analysis has some limitations. First, using semi-structured phone interviews to obtain data has some intrinsic limitations. The linguistic diversity of participants, for example, was limited. The primary interviewer's ability to speak only two languages with fluency, French and English, disqualified potential participants who were not fluent in either, although one Spanish-language interview was accommodated. Furthermore, unspoken insights that may have been gleaned through the interviews were missed because they were not conducted face-to-face. The reliability and affordability of phone and Skype interviews (see [40]) outweighed these concerns as conducting interviews using this medium is what enabled us to recruit such an international sample of participants.

Second, although interviewing IPCs is useful, it results in a second-hand perspective on medical tourists' informal caregivers. We believe that the strengths of interviewing this group in the context of this study, particularly that they were able to comment on trends across many caregivers, outweighed this limitation. Every interview we conducted captured experiences with hundreds of caregivers, which was necessary to create the broad understanding that we sought to provide with this analysis. However, our analysis needs to be complemented with first-hand accounts from caregivers, and we intend on pursuing this in the next phases of our research. We believe that the limitations we have identified here are acceptable and do not significantly compromise the rigor of the study or analysis.

6. FUTURE RESEARCH DIRECTIONS

Although we provide the first dedicated investigation of the roles of informal caregivers in medical tourism, this article contributes only a small part to our understanding of them. We have not measured the frequency with which each role is adopted, we have not assessed the spatiality or temporality of these roles, nor have we considered the effectiveness of these roles in ameliorating or maintaining patient health or wellness or even offsetting what is known as 'caregiver burden'. These all serve as important directions for future research. For example, the collective impact of the stressors encountered by informal caregivers in the practice of care is referred to as caregiver burden [41, 42, 43]. Significant burden might lead to 'caregiver burnout' [41, 44, 45, 46]. In this article, we show that the friends and family members who accompany medical tourists abroad are indeed filling informal caregiving roles and it would be valuable to determine if and how the activities they undertake in their capacities as knowledge brokers, companions, and navigators result in exposure to burden and ultimately burnout. This knowledge could then be used to assist in identifying interventions to offset that burden if it is found to exist. The caregiver group could benefit from, for example, the development of informational tools that anticipate and address the possibility of caregiver burnout. Although there has been some discussion of the need to create reliable, evidence-informed informational tools about medical tourism and enhance patients' access to such sources [47, 48], there has been no consideration of whether or not medical tourists' informal caregivers could similarly benefit from enhanced access to credible information. Second, any demonstration of burden or burnout among this group would contribute new evidence to the health equity debates that exist around the practice of medical tourism that focus on discerning who 'benefits' and who 'loses' from the existence of this health services trade (see, for example, [21, 49]). Although informal caregivers are not addressed in contemporary analyses of the health equity effects of medical tourism, they need to be incorporated into examinations of this subject.

An important area for future research pertains to our participant group: IPCs. An enhanced understanding of the scope and scale of IPCs' roles and responsibilities as a whole is very much needed. Such knowledge is critical given that many aspects of informal caregivers' roles hinge on the time, energy, and attentiveness of IPCs. Yet, there is not an adequate understanding of IPC training regarding interactions with caregivers, or the proportion of their time allocations or work tasks assigned to dealing with such interactions. Furthermore, we have shown that IPC and informal caregiver roles intersect at many points in a medical tourist's journey. The implications of these intersections for the roles assumed by each group are unknown, such as the benefits and drawbacks for the health and wellbeing of the medical tourist. Such research would not only be important in increasing the knowledge of caregiver roles and responsibilities, it would also expand the existing knowledge of relationships between other medical tourism stakeholder groups. Dedicated attention to IPC roles and responsibilities could, for example, be used to create a schematic that demonstrates the intersection of the roles of IPCs, formal health care providers, other staff from the medical facility and from hospitality services such as hotel concierges and airline representatives, caregivers, and medical tourists.

7. CONCLUSIONS

We found that the friends and family members who accompany medical tourists abroad engage in caregiving roles almost continuously while abroad. They facilitate and supplement the efforts of formal health care providers at medical tourism facilities to ensure the patient's health and wellbeing and shoulder some of the responsibilities that might otherwise be assigned to patients. In this way, caregivers can act as both amplifiers and buffers: they amplify the efforts of the facility staff through facilitation and supplementation while buffering medical tourists from stresses stemming from responsibilities, anxieties, and discomforts. Although there are parallels between some of our findings about caregiver roles and those shared in the existing caregiving literature, the unique transnational care context of informal caregiving in medical tourism reveals activities and responsibilities assigned to particular roles that are specific to this particular care practice.

We believe that the knowledge that has been gleaned about a heretofore neglected medical tourism stakeholder group, namely patients' informal caregivers, and the roles filled by members of this group provides valuable insight into how the medical tourism industry operates. Given the integral roles that friends and family members play (and particularly while they are abroad), the practice of medical tourism and thus the industry that supports it seems highly dependent on their unpaid care work. As with other informal caregivers, these individuals are effectively overlooked "shadow workers" [50] – unpaid, untrained, and largely unrecognized care providers - in what is often reported as a highly lucrative industry. We believe that adopting this critical perspective in light of the findings is essential to effectively address the question 'who ultimately benefits from medical tourism?' that is central to the health equity debates that surround the medical tourism industry. Although this question is far beyond the scope of this article, we do demonstrate the value of considering the roles that *every* stakeholder group plays in enabling this global health services practice. Without having such knowledge it becomes impossible to effectively determine the full scope of the health equity impacts of medical tourism, mitigate the negative ones, and enhance the positive ones, which is a key issue currently being discussed in research and policy circles [6, 14, 21,51, 52].

ENDNOTE

[a]Informal caregivers to patients recovering from cosmetic or bariatric surgeries were sometimes noted to be problematic to patients' wellbeing because they can become visibly distressed by a patient's appearance. Therefore, some IPCs who worked at facilities specializing in these procedures advised medical tourists to travel unaccompanied. See [53] for further examination.

ACKNOWLEDGEMENTS

We would like to thank those who participated in this study for sharing their experiences and views with us. Thank you to Alejandro Ceron for conducting the Spanish language interview. The study was funded by an Operating Grant from the Canadian Institutes of Health Research (funding reference #115692). VAC is also funded by a Scholar Award from the Michael Smith Foundation for Health Research.

AUTHORS' CONTRIBUTIONS

VC conducted the interviews, led the coding, and played a leadership role in writing this article. VAC is lead investigator on the study that funded this research and developed the protocol. She contributed to all aspects of the study and was heavily involved in preparing this manuscript. JS and LT contributed input to the data collection instrument, participated actively in the face-to face analysis meeting, and provided feedback on drafts of this paper. JS also aided in recruitment. All authors reviewed and approved the submitted manuscript.

REFERENCES

1. Martin R: Clark Howard Discusses His Cancer Diagnosis With Dr. Randy Martin. 2009, Atlanta: WSB-TV,http://www.wsbtv.com/news/news/clark-howard-discusses-his-cancer-diagnosis-with-d/nFBhr/,

2. Smeltzer SC, Bare BG, Hinkle JL, Cheever KH: Health care delivery and nursing practice. Brunner and Suddarth's Textbook of Medical-Surgical Nursing: In One Volume. 2010, Philadelphia: Lippincott Williams & Wilkins, 9-10. 12

3. Schein C, Gagnon AJ, Chan L: The association between specific nurse case management interventions and elder health. J Am Geriatr Soc. 2005, 53: 597-602.

4. Schoenborn NL, Arbaje AI, Eubank KJ, Maynor K, Carrese JA: Clinician roles and responsibilities during care transitions of older adults. J Am Geriatr Soc. 2013, 61: 231-236.

5. GG Van Bruchem-van de S, van der Arend AJG, Huijer Abu-Saad H, Spreeuwenberg C, van Wijmen FCB, ter Meulen RHJ: The role of nurses in euthanasia and physician-assisted suicide in The Netherlands. J Med Ethics. 2008, 34: 254-258.

6. Connor S, Adams J: Caregiving at the end of life. Hastings Cent Rep. 2003, 32 (Suppl 2): S8-S12.

7. Karimli L, Ssewamala FM, Ismayilova L: Extended families and perceived caregiver support to AIDS orphans in Rakai district of Uganda. Child Youth Serv Rev. 2012, 34: 1351-1358.

8. Ussher JM, Wong WKT, Perz J: A qualitative analysis of changes in relationship dynamics and roles between people with cancer and their primary informal carer. Health. 2011, 15: 650-667.

9. Halvorsen K, Førde R, Nortvedt P: The principle of justice in patient priorities in the intensive care unit: the role of significant others. J Med Ethics. 2009, 35: 483-487.

10. Lam RC: Contradictions between traditional Chinese values and the actual performance: a study of the caregiving roles of the modern sandwich generation in Hong Kong. J Comp Fam Stud. 2006, 37: 299-313.

11. Sinding C: Informal care – two tiered care? the work of family members and friends in hospitals and cancer centres. J Sociol Soc Welfare. 2004, 31: 69-86.

12. Connell J: Contemporary medical tourism: conceptualisation, culture and commodification. Tourism Manage. 2013, 34: 1-13.

13. Ehrbeck T, Guevara C, Mango PD, Cordina R, Singhal S: Health care and the consumer. McKinsey Quart. 2008, 4: 80-91.

14. Lunt N, Smith R, Exworthy M, Green ST, Horsfall D, Mannion R: Medical tourism: treatments, markets and health system implications: a scoping review. 2012, Geneva: OECD Directorate for Employment, Labour and Social Affairs,http://www.oecd.org/els/health-systems/48723982.pdf,

15. Deloitte Center for Health Solutions: Medical tourism consumers in search of value. 2008, New York: Deloitte,http://www.deloitte.com/

assets/Dcom-unitedStates/Local%20Assets/Documents/us_chs_MedicalTourismStudy(3).pdf

16. NaRanong A, NaRanong V: The effects of medical tourism: Thailand's experience. Bull World Health Organ. 2011, 89: 336-344.

17. Pocock NS, Phua KH: Medical tourism and policy implications for health systems: a conceptual framework from a comparative study of Thailand, Singapore and Malaysia. Global Health. 2011

18. Tourism Research and Marketing: Medical tourism: a global analysis. 2006, Arnhem: Association for Tourism and Leisure Education,http://www.atlas-webshop.org/Medical-tourism-a-global-analysis,

19. Alleman BW, Luger T, Schacht Reisinger H, Martin R, Horowitz MD, Cram P: Medical tourism services available to residents of the United States. J Gen Intern Med. 2011, 26: 492-497.

20. Burkett L: Medical tourism: concerns, benefits, and the American legal perspective. J Legal Med. 2007, 28: 223-245.

21. Johnston R, Crooks VA, Snyder J, Kingsbury P: What is known about the effects of medical tourism in destination and departure countries? a scoping review. Int J Equity Health. 2010, 9: 24-10

22. Kangas B: Hope from abroad in the international medical travel of Yemeni patients. Anthropol Med. 2007, 14: 293-305.

23. Yeoh E, Othman K, Ahmad H: Tourism management understanding medical tourists: word-of-mouth and viral marketing as potent marketing tools. Tourism Manage. 2013, 34: 196-201.

24. Yu JY, Ko TG: A cross-cultural study of perceptions of medical tourism among Chinese, Japanese and Korean tourists in Korea. Tourism Manage. 2012, 33: 80-88.

25. Bevan JL, Pecchioni LL: Understanding the impact of family caregiver cancer literacy on patient health outcomes. Patient Educ Couns. 2008, 71: 356-364.

26. Molloy GJ, Johnston DW, Witham MD: Family caregiving and congestive heart failure: review and analysis. Eur J Heart Fail. 2005, 7: 592-603.

27. Grace M: State of the Heart: A Medical Tourist's True Story of Lifesaving Surgery in India. 2007, Oakland: New Harbinger Publications

28. Rose DA: Larry's Kidney: Being the True Story of How I Found Myself in China with My Black Sheep Cousin and His Mail-Order Bride, Skirting the Law to Get Him a Transplant–and Save His Life. 2009, New York: William Morrow

29. Fereday J, Muir-Cochrane E: Demonstrating rigor using thematic analysis: a hybrid approach of inductive and deductive coding and theme development. Int J Qual Method. 2006, 5: 80-92.

30. Braun V, Clarke V: Using thematic analysis in psychology. Qual Res Psychol. 2006, 3: 77-101.

31. Solomon H: Affective journeys: the emotional structuring of medical tourism in India. Anthropol Med. 2011, 18: 105-118.

32. Kangas B: The burden of pursuing treatment abroad: three stories of medical travelers in Yemen. Global Soc Pol. 2010, 10: 306-314.

33. Kingsbury P, Crooks VA, Snyder J, Johnston R, Adams K: Narratives of emotion and anxiety in medical tourism: on state of the heart and Larry's kidney. Soc Cult Geogr. 2012, 13: 361-378.

34. National Alliance for Caregiving, AARP: Caregiving in the U.S. 2004, Bethesda, Washington: National Alliance for Caregiving, AARP,http://www.caregiving.org/data/04finalreport.pdf,

35. Mezey M: Nurses and their changing relationship to family caregivers. The Cultures of Caregiving: Conflict and Common Ground Among Families, Health Professionals, and Policy Maker. Edited by: Levine C, Murray TH. 2004, Baltimore: The John Hopkins University Press, 70-86.

36. Neufeld A, Harrison M: Nursing and Family Caregiving: Social Support and Nonsupport. 2010, New York: Springer Publishing Company

37. Donelan K, Hill CA, Hoffman C, Scoles K, Hollander Feldman P, Levine C, Gould D: Challenged to care: informal caregivers in a changing health system. Health Affair. 2002, 21: 222-231.

38. Petronio S, Sargent J, Andea L, Reganis P, Cichocki D: Family and friends as healthcare advocates: dilemmas of confidentiality and privacy. J Soc Pers Relat. 2004, 21: 33-52.

39. Weinberg DB, Lusenhop RW, Hoffer Gittell J, Kautz CM: Coordination between formal providers and informal caregivers. Health Care Manage Rev. 2007, 32: 140-149.

40. Laws S, Harper C, Marcus R: Telephone interviews. Research for Development. 2003, Thousand Oaks, CA: SAGE, 298-299.

41. Lee H, Singh J: Appraisals, burnout and outcomes in informal caregiving. Asian Nurs Res. 2010, 4: 32-44.

42. Ray RA, Street AF: The dynamics of socio-connective trust within support networks accessed by informal caregivers. Health. 2011, 15: 137-152.

43. Roth DL, Perkins M, Wadley VG, Temple EM, Haley WE: Family caregiving and emotional strain: associations with quality of life in a large national sample of middle-aged and older adults. Qual Life Res. 2009, 18: 679-688.

44. Murphy NA, Christian B, Caplin DA, Young PC: The health of caregivers for children with disabilities: caregiver perspective. Child Care Health Dev. 2007, 33: 180-187.

45. Takai M, Takahashi M, Iwamitsu Y, Oishi S, Miyaoka H: Subjective experiences of family caregivers of patients with dementia as predictive factors of quality of life. Psychogeriatr. 2011, 11: 98-104.

46. Yılmaz A, Turan E, Gundogar D: Predictors of burnout in the family caregivers of Alzheimer's disease: evidence from Turkey. Australas J Age. 2009, 28: 16-21.

47. Lunt N, Hardey M, Mannion R: Nip, tuck and click: medical tourism and the emergence of web-based health information. Med Inform J. 2010, 4: 1-11.

48. Penney K, Snyder J, Crooks VA, Johnston R: Risk communication and informed consent in the medical tourism industry: a thematic content analysis of Canadian broker websites. BMC Med Ethics. 2011, 12: 17-10.

49. Chen YYB, Flood CM: Medical tourism's impact on health care equity and access in low- and middle-income countries: making the case for regulation. J Law Med Ethics. 2013, 41: 286-300.

50. Armstrong P, Armstrong H, Coburn D: Unhealthy Times: Political Economy Perspectives on Health and Care. 2001, London: Oxford University Press

51. Forgione DA, Smith PC: Medical tourism and its impact on the US health care system. J Health Care Financ. 2007, 34: 27-35.

52. Hopkins L, Labonté R, Runnels V, Packer C: Medical tourism today: what is the state of existing knowledge?. J Public Health Pol. 2010, 31: 185-198.

53. Casey V, Crooks VA, Snyder J, Turner L: "You're dealing with an emotionally charged individual...": an industry perspective on challenges posed by medical tourists' informal caregiver-companions. Global Health. 2013, 9: 31

Chapter 11

Risk communication and informed consent in the medical tourism industry: A thematic content analysis of canadian broker websites

Kali Penney[1], Jeremy Snyder[2*], Valorie A Crooks[3] and Rory Johnston[3]

[1]Faculty of Medicine, University of Calgary, 2500 University Drive, NW, Calgary, Alberta, Canada.

[2]Faculty of Health Sciences, Simon Fraser University, 8888 University Drive, Burnaby, British Columbia, Canada.

[3]Department of Geography, Simon Fraser University, 8888 University Drive, Burnaby, British Columbia, Canada.

ABSTRACT

Background

Medical tourism, thought of as patients seeking non-emergency medical care outside of their home countries, is a growing industry worldwide. Canadians are amongst those engaging in medical tourism, and many are helped in the process of accessing care abroad by medical tourism brokers - agents who specialize in making international medical care arrangements for patients. As a key source of information for these patients, brokers are likely to play an important role in communicating the risks and benefits of undergoing surgery or other procedures abroad to their clientele. This raises important ethical concerns regarding processes such as informed consent and the liability of brokers in the event that complications arise from procedures. The purpose of this article is to examine the language, information, and online marketing of Canadian medical tourism brokers' websites in light of such ethical concerns.

Methods

An exhaustive online search using multiple search engines and keywords was performed to compile a comprehensive directory of English-language Canadian medical tourism brokerage websites. These websites were examined using thematic content analysis, which included identifying informational themes, generating frequency counts of these themes, and comparing trends in these counts to the established literature.

Results

Seventeen websites were identified for inclusion in this study. It was found that Canadian medical tourism broker websites varied widely in scope, content, professionalism and depth of information. Three themes emerged from the thematic content analysis: training and accreditation, risk communication, and business dimensions. Third party accreditation bodies of debatable regulatory value were regularly mentioned on the reviewed websites, and discussion of surgical risk was absent on 47% of the websites reviewed, with limited discussion of risk on the remaining ones. Terminology describing brokers' roles was somewhat inconsistent across the websites. Finally, brokers' roles in follow up care, their prices, and the speed of surgery were the most commonly included business dimensions on the reviewed websites.

Conclusion

Canadian medical tourism brokers currently lack a common standard of care and accreditation, and are widely lacking in providing adequate risk communication for potential medical tourists. This has implications for the informed consent and consequent safety of Canadian medical tourists.

1. BACKGROUND

Medical tourism is a practice whereby patients travel outside of their home country in order to receive medical care. The medical tourism industry is perceived as rapidly expanding, though precise numbers on patient flows are unknown [1]. For the purposes of this paper, we understand medical tourism to entail patients traveling internationally outside of established cross-border care arrangements with accessing medical care as their primary goal, which is usually paid for out of pocket [1]. Patients may decide to pursue medical treatment abroad for any number of reasons, including a desire to gain access to treatments that are not approved or are otherwise available at home, the cost of treatment, and the length of treatment wait times [2]. With smaller, elective procedures it may be the draw of the destination itself or the anonymity provided by going abroad that motivates patient travel [3].

Recently, there have been widely publicized reports in the popular press about medical tourists dying or suffering complications from their surgeries [4, 5, 6]. In late 2010, Mahir Mostic, a 35 year old man from Ontario, Canada traveled to Costa Rica to receive the controversial 'liberation therapy' for multiple sclerosis and suffered complications from the surgery. He returned to Costa Rica for follow up care but subsequently died [4]. A woman from Alberta, Canada traveled abroad for breast augmentation in 2010, and returned home with a serious infection that resulted in the eventual loss of the majority of her breast tissue [5]. Cases like these illustrate the possible negative outcomes faced by medical tourists. Any surgery carries with it a certain degree of risk; there is always the possibility of anaesthesia complications, blood clots, uncontrolled bleeding and human error during surgery as well as delayed healing and infection after surgery [1]. Medical tourism, however, carries with it additional risks, including communication gaps between the physician and patient, taking lengthy flights post-operatively, exposure to medical malpractice abroad, difficulties obtaining follow up care, and the danger of infectious disease transmission, among others [1].

A highly visible and influential resource available to patients thinking about going abroad for medical care is the medical tourism broker. For the purpose of this paper a broker is defined as a third party who connects patients to hospitals or doctors in another country. Brokers play a unique role in the practice of medical tourism, and yet remain largely unregulated and poorly understood. The exact percentage of medical tourists who choose to use these brokers' services is unknown, but it is thought that they are likely playing a significant role in medical tourists' health decision-making [7]. For many medical tourists, a broker provides critical information and services that would be difficult to acquire independently, particularly when traveling to a foreign country for the first time. Brokers can help make travel arrangements, suggest physicians and facilities abroad, book surgeries, assist in the transportation and translation of medical records, and help arrange for follow up care and oversee postoperative complications [8, 9].

The internet has made it possible for more people than ever before to explore medical treatment options abroad. The vast majority of brokers have a strong online presence and their websites are an important tool for patients seeking information about possible treatments and destinations [10] as well as financing and insurance information [11]. These websites are often the first source of information that patients are exposed to [1, 7, 11, 12]. Therefore, the information about the benefits and risks of engaging in medical tourism presented on these sites may have a strong impact on whether or not and how the prospective medical tourist pursues treatment abroad [12]. While medical tourists working with a broker will typically be given additional information beyond what is available online [6], brokers' websites likely help form the patient's first impression of medical tourism.

Given the risks associated with surgery in general and medical tourism in particular, it is imperative from the standpoint of medical ethics to assess how patients understand these risks and how they are communicated to them by others, especially when coupled with the increasing visibility of medical tourism as a global health services practice. Informed consent is a central ethical value in medical practice [13]. What currently remains unknown is whether medical tourism broker's websites serve to enhance or undermine the capacity of potential medical tourists to make informed choices about engaging in this practice. While potential medical tourists can avail themselves of other sources of information [1, 7], the central role of broker websites in forming initial impressions of the safety of medical tourism means that any failure of these sites to communicate risks effectively would raise serious ethical concerns [14]. In this article we examine Canadian medical tourism brokers' involvement in risk communication to prospective patients via their websites in order to address this knowledge gap. We accomplish this through conducting a thematic content analysis of 17 English-language brokerage websites. We seek specifically to analyze the way brokers communicate risk via their promotional websites and the potential impact of their websites on informed consent for patients, which contributes to the nascent literature examining brokers' roles and responsibilities in medical tourism [6, 7, 11, 12, 14].

2. RISK AND INFORMED CONSENT IN MEDICAL TOURISM

Effective communication is essential to medical practice, and medical ethicists emphasize the importance of informed consent as a dimension of effective communication [15]. Informed consent carries with it two levels of moral responsibility on the part of the health professional: adequate information must be communicated to the patient and the patient must understand the information they receive in order to make an informed decision regarding treatment [13]. This is particularly important when considering a medical procedure that is inherently risky or poorly understood [16]. Because the possibility of a negative treatment outcome is appealing neither to the physician nor the patient, there is a tendency to avoid direct discussion of risk, which can lead to patients making decisions without having truly considered the risks involved [17]. Patient assessment of risk is also subject to various biases that can undermine the clear communication of these risks, making it important to factor these biases into risk communication strategies [18].

Medical tourism expands the range of individuals giving patients health advice with the addition of MT brokers and foreign care providers [1], thereby further complicating the communication of risk to these patients and thus ultimately compromising their ability to achieve informed consent. Information about the quality of care available abroad and the possible risks involved in a procedure is important for the patient to have and to understand [1]. While under normal circumstances patients may receive this information from their family doctor or from a nurse or surgeon at a hospital in their home country [18], medical tourists may need to rely on additional sources of information [1, 19]. As medical tourism involves travel to a foreign country that patients and their families may be unfamiliar with, this practice creates uncertainties not found in traditional treatment that may add additional risks and informational burdens [20]. Moreover, the choice to travel abroad may strain relationships with the patient's family doctor and other local medical professionals, further cutting off normal means of gaining medical advice [6].

Not only may risk communication be particularly complicated in the case of medical tourism, but the nature of the risks that patients may be exposed to can also be heightened. Long air travel has been shown to increase a person's risk of deep-vein thrombosis (DVT) [21] and it is suspected that this risk is further increased by having had recent surgery [22]. Communication between medical tourists and health care professionals abroad may be difficult if there is a language barrier, thus limiting risk-related communication [23]. The quality of a hospital or of a physician's training may be difficult to ascertain if the credentials are foreign to the patient, further complicating patient understanding of risk [24]. In the case of a physician's error or hospital negligence, it is often difficult to receive compensation for malpractice in many countries [25]. Follow up care for medical tourists is uniquely challenging as they will be returning to their home country and may have difficulty contacting the physician or surgical team abroad should complications or questions arise post-operatively, which can pose as a post-operative health risk [14]. These potential risk factors have been previously identified, and many of them are preventable.

3. METHODS

To identify English-language Canadian medical tourism brokerage websites for inclusion in this analysis, we first compiled a listing of company names mentioned in published media accounts of medical tourism and reviewed for Canadian membership in the Medical Tourism Association (MTA). Following this we searched the identified company names in an internet search engine to locate their websites. Next we ran a series searches on two internet search engines, exhaustively searching combinations of the keywords Canad*, medical, tour*, and surger* using Boolean operators. Twenty two brokerages were initially identified using these strategies in summer of 2010. In reviewing their websites, five were excluded from the study as they appeared to be inactive (e.g., no current contact information), not focused on medical tourism (e.g., were selling insurance packages), or were not based in or operating out of Canada.

Following identification of the websites for inclusion in this analysis we performed a thematic content analysis. Thematic content analysis involves generating frequency counts of the dominant themes in a dataset that can be used to guide a thematic approach to analysis, which typically involves identifying themes within a dataset and comparing those themes to the study purpose and existing literature [26]. In keeping with this approach, the websites were first reviewed in order to develop a list of informational points that were commonly included. Additional informational points were added to the listing based upon our focus on risk communication and informed consent. Frequency counts of these informational points were generated, which formed the basis of the content component of our analysis. In reviewing the trends among the quantitative counts and comparing these trends to the existing literatures on medical tourism and informed consent, three themes were identified for the thematic component of the analysis. Following identification of these themes, consensus was sought among the team members to ensure confirmation of this interpretation. Data relevant to each of these themes were then compiled from the websites included in the study and reviewed again in order to focus the analysis.

4. RESULTS

A summary of the websites included in this analysis can be found in Table 1. A summary of the counts generated for the thematic content analysis is included in Table 2. In reviewing the content of the 17 websites included in this analysis, we found that Canadian medical tourism broker websites varied widely in scope, content, professionalism, and depth of information. Three themes emerged from the thematic content analysis: training and accreditation, risk communication, and business dimensions. In this section we examine each separately, identifying commonalities and differences across the reviewed websites. We provide direct quotations from the reviewed websites to illustrate the points raised in this section, and provide a number beside each quote (e.g., #2) that corresponds to 'company' column in Table 1.

4.1 Training & Accreditation

Three types of accreditation and/or training were mentioned explicitly on brokers' websites: (1) accreditation of international hospitals; (2) training and accreditation of international physicians; and (3) brokers' own training and accreditation. There was a certain degree of continuity between the three different types of training and accreditation identified; 29.4% of websites did not mention any specific credentials for the hospitals or physicians they referred to nor did they display logos of membership for their own brokerage. The remaining 70.6% varied in their level of professional accreditation and communication of physicians' qualifications.

Table 1. Medical Tourism Websites Included in Content Analysis

Company	Website	Province of Headquarters
1	http://timelymedical.ca	British Columbia
2	http://www.meditours.org	British Columbia
3	http://www.hsi-ssi.com/en	Quebec
4	http://www.surgicaltourism.ca	British Columbia
5	http://ecumedical.com/ecom.asp?*	Ontario
6	www.healthvacations.ca	None noted
7	http://www.debsonmedicaltourism.com	Quebec
8	http://www.medextra.com	Quebec
9	www.gosculptura.com	Quebec
10	http://ihcproviders.com	Ontario
11	http://passportmedical.com	British Columbia
12	http://www.intermedglobal.com	None noted
13	http://www.medipassionhealing.com*	Quebec
14	http://www.overseasmedicalservices.com*	Alberta
15	http://www.surgicalescape.com	Alberta
16	http://www.choicemedicalservices.com/	Manitoba
17	http://www.medasia.ca/	None noted

*Website used for study has either changed or been shut down since September 2010

Table 2. Informational points contained on Canadian medical tourism brokers' websites (N = 17)

Informational Points	number of websites including informational point	number of websites not including informational point
Evidence of professional membership or certification	6	11
Information on foreign physicians (e.g., certifications, training)	5	12
Explicit reference to foreign hospital accreditations	9	8
Explicit reference to foreign hospitals with JCI accreditation	6	11
Explicit reference to "medical tourism"	12	5
Explicit reference to "broker"	0	17
Explicit reference to "facilitator"	7	10
Tourism services offered (e.g., arranging sightseeing)	10	7
Images of tourist attractions	7	10
Health risks (general - e.g., surgical risks)	9	8
Health risks (specific to medical tourism - e.g., post-operative flight)	2	15
Patient testimonials	11	6
Pricing details	6	11

The current gold standard for international hospitals is Joint Commission International (JCI) accreditation [14], and four of the websites stated specifically that they only work with hospitals that are JCI accredited, with one explaining: "JCI Accredited hospitals and clinics must have the same or exceed standards compared to the United States in clinical outcomes, infection rates, mortality rates and cleanliness." The majority of websites made no mention of any type of specific hospital accreditation or certification, often using vague descriptions as a substitute: "all rooms are equipped with the very latest American and European technology" (#9); "Our standards are as high, and can exceed, those in the United States and United Kingdom" (#13); the hospitals use the "best technology available on the market today" (#2). These examples show the language that was commonly used to advertise services, with superlative descriptors of the quality of service and availability of new technologies appearing on all websites.

In addition to indicators of the quality and accreditation of foreign medical facilities, websites provided information on the quality of the personnel in these facilities through reference to their medical credentials. Medical credentials of foreign physicians were mentioned on 41.2% of websites, sometimes in a generalized way: "many of our doctors have been trained in the United States and the United Kingdom, right alongside the physicians and surgeons that are working in the city in which you live" (#13). Overall 52.9% of websites stated that the company worked with international physicians trained, at least in part, in either North America or Europe. Some websites not reporting credentials and training used other tactics to gain the trust of potential medical tourists: "The life expectancy in Cuba is equal to that found in developed countries... This type of statistic can only be achieved if the medical training and procedures are similar to those of developed countries" (#16).

While there was no single, dominant professional organization or association to which the brokers belonged, 35.3% of the websites featured symbols and logos of some type on their main page, usually representing membership in various organizations (e.g., MTA, Better Business Bureau, International Medical Travel Association, Medical Travel and Health Quality Alliance). Demonstrations of certification, rather than simply membership, on the websites were rare, with only one website displaying the logo of the MTA certification for brokers. It is worth noting that the MTA membership logo is extremely similar to the certification logo (the caduceus) with the exception of the word "member" instead of "certification," a distinction which may not be apparent or have meaning to website visitors. One other site included a symbol for an accrediting agency, the Medical Travel and Health Tourism Quality Alliance, which certifies brokers to be either a Certified Medical Concierge or a Certified International Patient Manager.

4.2 Risk Communication

A detailed search for any mention of risk on the reviewed websites revealed that 47.1% made no general or specific mention of risks at all. Of the 9 websites that did address risk, only two mentioned any specific risks associated with medical tourism, and these risks were mentioned only in order to illustrate how, by using a broker, they would be eliminated. Two sites mentioned that there was a small risk involved in obtaining a medical procedure, but were careful to separate that risk from any risks associated with traveling by reminding the patient that the same risk was involved with the medical procedure if done in the home country.

Risk of DVT due to post-operative travel was mentioned on two websites, but only in a context that implied that neither company's patients would have an increased risk of DVT due to their short flights. For example, a brokerage that only arranges for medical tourism between Canada and the United States lists several risks of traveling to a country outside of North America for surgery, including DVT due to long-haul flights. Their website provides a good example of how careful wording was often used on when discussing risk: "In the unlikely event of a complication during surgery, which is unrelated to the reason for traveling to the United States for medical care, the hospital will stabilize the patient and send him/her back to Canada by ambulance" (#1). Another example of careful wording is: "As with any other surgical procedure, there are risks and complications involved which are always measured. Certain risks depend on the individual surgeon and are avoided without being acknowledged due to the surgeon's expertise" (#9).

Discussion of negative outcomes and risks occurred most often on the Frequently Asked Question or Disclaimer sections of websites, and only 17.6% addressed possible negative outcomes. Such discussions appeared to be aimed at reassuring potential clients, as shown in this example: "As with any medical procedure at home or away, there is always a small chance of a complication, so one should always plan for this eventuality. That is why medical complication insurance is available...so that you can have peace of mind" (#11). While most websites provided minimal overviews of the possible negative outcomes to medical tourism, the majority had many examples of positive outcomes. One way this was done was through the inclusion of patient testimonials, where 70.6% of the websites contained at least one personal statement from a former patient or from a physician abroad. One company promised to put interested clients personally in touch with former clients for a live testimonial experience.

4.3 Business Dimensions

Not surprisingly, websites conveyed information about the broker's business to potential medical tourists. One of the business dimensions covered pertained to brokers' involvement in follow up care, with 52.9% of the websites promising quality follow-up care upon return home. While those that did not make this promise typically did not address follow up care at all, one explicitly stated that the brokerage had no responsibility in overseeing such care, stating that: "once arrangements have been made to the mutual satisfaction of our client and their physician, we have no ongoing involvement in the doctor-patient relationship" (#1). For the brokers who were involved in arranging for it, follow up care services varied between companies. These services included arranging phone calls between the patient and the specialist abroad, having a report sent to the home physician, organizing rehabilitation, telehealth consultations, and answering questions. It was often unclear whether these services would come at an additional cost to the patient.

Medical tourism is often associated with price savings for patients. Nonetheless, only 29.4% of the websites included specific prices for various treatments. When sites did not list their prices, they often included percentages representing the average savings a tourist could expect. For example, one website stated that "if you have received quotes in America for $10,000, you can expect the same procedure to cost between $2,500 and $5,000" (#13) while another noted that "care comes at a cost that is 40-80% lower than the cost of the same procedures in the United States" (#6). It is interesting to note that United States dollars were quoted in these examples despite the fact that these companies were operating in Canada, although references to faster wait times for procedures typically quoted Canadian averages. For example, a website listed their services as being "4-10 times faster than in Canada" (#6).

The large majority (70.6%) of the brokers branded themselves as being involved in the business of 'medical tourism' via their websites. Brokers that did not use the medical tourism label instead used labels such as: medical travel, healthcare management, private pay medical services, and travel for health services. This divide in branding or labelling extended to the images used on sites, where medical tourism companies commonly depicted images of tropical flowers and palm trees, images of celebrations and festivals that occur in destination countries, and pictures of beach and ocean views, while sites not adopting the medical tourism label had little or no tourism-related imagery and were thus not marketing the tourism dimensions of international patient travel.

The term 'broker' or 'brokerage' was not used at all in the sample of reviewed websites. Meanwhile, this term is widely used in the academic and popular literature regarding medical tourism when characterizing these businesses and the services they offer [27, 28, 29]. The term that appeared most frequently (47.1%) was 'facilitator', and it appeared in a variety of permutations, including 'medical travel facilitator', 'health-medical tourism facilitator', and 'medical tourism facilitator.' Other descriptors included 'medical tourism service provider', 'referral agency', 'medical tourism advocates', 'international patient support' and 'medical tourism guidance.'

5. DISCUSSION

Medical tourism brokers play an important role in the medical tourism industry and have helped make traveling abroad for medical care a viable option for more people than ever before [6, 7, 11,12, 14]. Through our assessment of 17 Canadian brokers' websites we have been able to gain an understanding of the quantity and quality of online information available to persons using Canadian brokers. Their websites were clearly designed to sell brokers' services to potential medical tourists. For this reason, the information provided through these websites is geared more toward promotional ends rather than helping to inform patients of the full range of their treatment options or communicating the potential risks inherent in medical tourism. Mason and Wright identified a similar trend in their analysis of 66 international medical tourism websites (which included broker and non-broker sites) [12]. Crooks et al.'s thematic content analysis of promotional brochures from medical tourism hospitals in India also revealed a strong emphasis on characterizing credentials and accreditation, services, and specializations over issues such as cost and follow up care

[30]. When the findings of our analysis are considered in relation to these other studies it becomes clear that a lack of emphasis on risk communication in medical tourism marketing materials is not specific to Canadian brokers but, instead, is more likely an industry-wide phenomenon.

The findings shared above highlight many issues that hold significant implications for Canadian patients' informed consent when engaging in medical tourism, four of which we expand upon in this section. First, our review revealed that only 2 of the 17 brokerages had been certified by an accrediting body, while many websites had various logos signifying membership with the MTA and other groups. While these logos are meant to assure the patient of the broker's professional status, they do not typically guarantee any particular standard of operation. The risk generated by this lack of standardization is that it allows brokers with lower standards and who do less to mitigate and inform patients of risks to operate amidst competent brokers, with little means of differentiating between the two [6, 14]. This lack of clarity may compromise a patient's ability to make informed decisions regarding their treatment, thereby ultimately compromising informed consent, and puts them in a position where they may be unsure of the consequences of deciding to use a particular broker to book a procedure abroad.

A second issue related to informed consent emerging from the findings pertains to the role of terminology in clarifying what services patients are purchasing. More specifically, the terminology chosen by the brokers may factor into how a patient views their services. The term 'medical tourism' has been described as that which is most commonly used within the industry and by the media [31], and the majority of the brokers in this study adopted it on their websites. Bioethicists and other scholars have observed that the term itself may be problematic; the word 'tourism' evokes images of beaches and sightseeing, potentially undermining the seriousness of the medical procedure in the patient's mind [19, 31, 32]. As such, the term may under-represent the risks involved in going abroad for medical care. Related to this is a need to develop standardized role terminology. The websites we reviewed uniformly avoided the term 'broker' that is so oft used in popularized contexts. This is likely deliberate; the term 'facilitator' evokes the image of a helper, someone making life easier, while the term 'broker' brings to mind the world of business and money-making. Moreover, a 'facilitator' may be seen more to have the patient's interests at heart, thus reducing risks, while a 'broker' might be seen as engaging in more mercenary, risk-taking behaviour, making the broker's perceived role an important element in the communication and understanding of risk.

The messages conveyed by the Canadian medical tourism broker websites we reviewed were overwhelmingly positive, which raises a third issue regarding informed consent. Positive, if not glowing, patient testimonials were more frequently present than not, while acknowledgement of risks were notably rare. This finding is consistent with those of other studies that have examined promotional information in medical tourism and have ultimately questioned the trustworthiness of what is being reported and its ability to support patient decision-making [7, 14, 30]. While it is understandable that medical tourism brokers, in an effort to portray their product in the most favourable light possible, may be reluctant to address the issue of risk on their websites, this omission is potentially misleading and unethical. Brokers may be inclined to leave the discussion of risk until later in the process of arranging care abroad or until they meet with the patient in person, but the website is likely the first and possibly main source of information for many patients [7]. In order to make a truly informed decision about whether or not to proceed with arrangements - and thus ultimately to give informed consent for medical care abroad - they ought to have access to balanced and accurate risk information at the beginning of their decision making process.

The final issue related to informed consent emerging from the findings of our thematic content analysis relates to information, or the lack thereof, about follow up care on the reviewed websites. Generally, follow up care details were rarely included in brokers' websites. This is concerning because when Canadians undergo surgery domestically, follow up care is commonly coordinated between the hospital and the family physician and is generally organized on behalf of the patient. Medical tourism disrupts this standard practice [33, 34]. Medical tourists will inevitably be moving between two different health care systems that are not connected, making it unclear exactly who should be responsible for organizing check-ups and rehabilitation after the procedure [6]. If follow up care details are not shared with the patient initially, or if it is not noted explicitly that patients are expected to arrange for their own follow up care upon return home, they may decide to proceed with purchasing medical care abroad while being unaware of the amount of care needed once they have returned home. This poses a risk to medical tourists on many fronts. It may negatively impact their health should they not be in a position to navigate their home health care system upon return home without the support or referral of a domestic surgeon [1]. It may also lessen the quality of care patients receive upon returning home in that they may underestimate the informational needs of service providers thereby compromising care quality. One way to address these concerns is to encourage patients to communicate their plans to engage in medical tourism to their family doctors before going abroad as this may lead to consideration of follow up care before departure [30]. Such patients will have a broader range of sources from which they are gathering information and will be able to come to a more informed decision on whether or not they want to consent to the procedure.

As we have noted, ethical and legal debates about the regulation of the medical tourism industry exist in the scholarly literature [14, 35]. In our own research we have found that Canadian medical tourism brokers believe they have significant roles to play not only within the industry but to medical tourists specifically, particularly to ensure these patients' safety [6]. The provision of informed consent falls naturally within the roles they espouse toward patients, but brokers' perspectives on this needs to be sought in order to ascertain whether or not they agree in both principle and practice.

The Canadian government is involved in protecting Canadian patients and ensuring their health through the public health care system, and as such is well positioned to address some of the issues that hold implications for medical tourists' informed consent that we have identified in this analysis. As brokers are involved in patient care, directly or indirectly, and have a unique relationship with the medical tourist, it would seem fitting that the government take steps to monitor their work and ensure they are meeting standards of care for patients. Turner has made a broad call for regulatory involvement by the American government in regard to medical tourism brokerage companies operating in the United States [14]. While the insights gained from our study of broker websites does not support the full range of regulations suggested by Turner, some of which would not protect patient choice (e.g., restricting brokers from advertising services not approved in their local jurisdictions), certain areas of intervention are justified by our findings. More specifically, Canadian government intervention in the areas of medical tourism brokers' involvement in follow up care and risk communication seems warranted given the lack of apparent consideration given to these domains by brokers currently as assessed by our review of their websites. Such intervention need not be complex, and could include developing informational resources for Canadian medical tourism brokers or requiring brokers to disclose risk information on their websites. In general, government regulators have an obligation to oversee and set standards for the information available to potential medical tourists in order to ensure that the risks of this practice are understood and that false, misleading, or unverifiable claims are not made by brokers that would unethically undermine patient informed consent [36].

6. LIMITATIONS

One limitation is that the information we used to draw conclusions about brokers' business practices was limited to that which was found on their websites. We are aware that many brokers use brochures, information booklets, e-mails, and face-to-face conversations in addition to their websites as means of disseminating information to their clients; these other forms of communication may provide valuable information not found on the websites. That being said, we do believe that the websites are a key source of information and that they play a significant role in patients' decision-making processes and formation of initial impressions when considering medical tourism and so warrant explicit consideration.

Another limitation is that the sample of websites reviewed may not be exhaustive of Canada-based brokers. We made all reasonable efforts to identify medical tourism brokerages based in Canada with English-language websites; however, it is possible that there are more companies currently operating in Canada than were identified in this study (e.g., those operating exclusively in French). Additionally, we did not include companies whose primary business was standard travel and tourism but who also had a medical tourism component. Nonetheless, we believe the included websites serve as a representative sample of medical tourism brokers in Canada and certainly those most accessible to potential medical tourists online. It is worth noting that three of the websites reviewed have been subsequently shut down or are under construction, an issue similarly encountered by Cormany and Baloglu in their analysis of international medical travel websites [29]. Our analysis, therefore, is of a representative snapshot of an evolving industry.

7. CONCLUSIONS

Our review of 17 Canadian medical tourism broker websites revealed that medical tourists may be exposed to incomplete or misleading information about the risks involved with this practice. Training and accreditation of these brokers was uneven and the symbols of professionalization found on these websites were more likely to confuse than enlighten potential medical tourists. The terminology used by brokers may obfuscate their roles, including the place of tourism activities in medical tourism. Broker websites were also found generally not to provide information on the risks of this practice and information on the brokers' business, including pricing and the provision of follow up care. These shortcomings are matters of grave ethical concern as they threaten the ability of patients to make informed decisions regarding their medical care.

Our findings support increased government oversight of brokers in the name of improving the communication of risks to patients and better informing their choice regarding whether or not to engage in medical tourism, thereby ultimately enhancing informed consent. Protecting patient choice justifies some previously called-for interventions, including government review of broker accreditors, oversight of the information provided on websites, and requirements for transparency regarding the provision of follow up care and legal liability [14, 27]. Other regulations are not justified by an interest in informed consent and may undermine patient choice. Creating support for the development of oversight mechanisms for the medical tourism industry will be aided by continued study of medical tourism brokers, including other country-specific studies like our own and reviews of the information provided by brokers beyond their publicly accessible websites.

ACKNOWLEDGEMENTS

We would like to thank Krystyna Adams for research support in the drafting of this manuscript and Victoria Casey for help in formatting references. This research was supported by funding from the Canadian Institutes of Health Research.

AUTHORS' CONTRIBUTIONS

KP wrote the background, results, discussion, and conclusion sections and gathered website data for this analysis. JS contributed to the drafting and editing of the background, results, discussion, and conclusion sections and to data analysis. VC wrote the methods section, edited the full manuscript, and contributed to data analysis. RJ aided in identifying the list of brokers, edited the full manuscript, and contributed to data analysis. All authors read and approved the final manuscript.

REFERENCES

1. Crooks VA, Kingsbury P, Snyder J, Johnston R: What is known about the patient's experience of medical tourism? a scoping review. BMC Health Serv Res. 2010

2. Abdullah BJJ: The sky is falling. Biomed Imaging Intervention J. 2006, 2: e29-

3. Connell J: Medical tourism: sea, sun, sand and... surgery. Tourism Manage. 2006, 27: 1093-1100.

4. Morrow A: Man dies after controversial MS treatment, doctor says. [http://www.theglobeandmail.com/news/national/man-dies-after-controversial-ms-treatment-doctor-says/article1805379/]

5. Jacobs M: Fixing foreign foul-ups. [http://www.edmontonsun.com/comment/columnists/mindelle_jacobs/2010/12/07/16465056.html]

6. Snyder J, Crooks VA, Kingsbury P, Adams K, Johnston R: The 'Patient's Physician One-Step Removed': The evolving roles of medical tourism facilitators. J Med Ethics. 2011, 37 (Suppl 9): 530-534.

7. Lunt N, Hardey M, Mannion R: Nip, tuck and click: medical tourism and the emergence of web-based health information. Open Med Info J. 2010, 4: 1-11.

8. Turner L: "Medical tourism" and the global marketplace in health services: U.S. patients, international hospitals, and the search for affordable health care. Int J Health Serv. 2010, 40: 443-67.

9. Klaus M: Outsourcing vital operations: What if US health care costs drive patients overseas for surgery?. Quinnipiac Health Law J. 2006, 9: 219-247.

10. Impicciatore P, Pandolfini C, Bonati M: Reliability of health information for the public on the world wide web: systematic survey of advice on managing fever in children at home. BMJ. 1997, 314: 1875-1879.

11. Cormany D, Baloglu S: Medical travel facilitator websites: an exploratory study of web page contents and services offered to the prospective medical tourist. Tourism Manage. 2010

12. Mason A, Wright KB: Framing medical tourism: An examination of appeal, risk, convalescence, accreditation, and interactivity in medical tourism websites. J Health Comm. 2011, 16 (Suppl 2): 1-15.

13. Beauchamp TL, Childress JF: Principles of Biomedical Ethics. 2001, New York: Oxford University Press, 5

14. Turner L: Quality in health care and globalization of health services: accreditation and regulatory oversight of medical tourism companies. Int J Qual Health Care. 2011, 23: 1-7.

15. Manson NC, O'Neill O: Rethinking informed consent in bioethics. 2007, New York: Cambridge University Press

16. Paling J: Strategies to help patients understand risks. BMJ. 2003, 327: 745

17. Ball LK, Evans G, Bostrom A: Risky business: challenges in vaccine risk communication. Pediatrics. 1998, 101: 453-458.

18. Bogardus ST, Holmboe E, Jekel JF: Perils, pitfalls, and possibilities in talking about medical risk. JAMA. 1999, 281: 1037-1041.

19. Kangas B: Hope from abroad in the international medical travel of Yemeni patients. Anthropol Med. 2007, 14: 293-305.

20. MacReady N: Developing countries court medical tourists. Lancet. 2007, 369: 1849-1850.

21. Ansari MT, Cheung MY, Huang JQ, Eklof B, Karlberg JPE: Traveler's thrombosis: a systematic review. J Travel Med. 2005, 12: 142-145.

22. Page SJ: The evolution of travel medicine: a new research agenda for tourism?. Tourism Manage. 2009, 30: 149-157.

23. Reed CM: Medical tourism. Med Clin NA. 2008, 92: 1433-1446.

24. Howze KS: Medical tourism: symptom or cure?. Ga L Rev. 2007, 41: 1013-1050.

25. Terry NP: Under-regulated health care phenomena in a flat world: medical tourism and outsourcing. West New Eng L Rev. 2007, 29: 421-72.

26. Weber RP: Basic Content Analysis. 1990, Newbury Park, CA: Sage, 2

27. Spece RG: Medical tourism: protecting patients from conflicts of interest in broker's fees paid by foreign providers. Ariz Legal Stud: Discussion Paper No. 09-41. 2009

28. Carrera P, Lunt N: A European perspective on medical tourism: the need for a knowledge base. Int J Health Serv. 2010, 40: 469-84.

29. Burkett L: Medical tourism: concerns, benefits, and the American legal perspective. J Legal Med. 2007, 28: 223-45.

30. Crooks VA, Turner L, Snyder J, Johnston R, Kingsbury P: Promoting medical tourism to India: messages, Images, and the marketing of international patient travel. Soc Sci Med. 2011, 72: 726-732.

31. Turner L: 'First world health care at third world prices': globalization, bioethics and medical tourism. BioSocieties. 2007, 2: 303-25.

32. Johnston R, Crooks VA, Snyder J, Kingsbury P: What is known about the effects of medical tourism in destination and departure countries? a scoping review. Int J Equity Health. 2010, 9: 24-10.

33. Turner L: Medical tourism: family medicine and international health-related travel. Can Fam Physician. 2007, 53: 1639-1641.

34. Crooks VA, Snyder J: Medical tourism: what Canadian family physicians need to know. Can Fam Physician. 2011, 57 (Suppl 5): 527-529.

35. Crooks VA, Snyder J: Regulating medical tourism. Lancet. 2010, 376 (Suppl 9751): 1465-1466.

36. Faden RR, Beauchamp TL, King NMP: A history and theory of informed consent. 1986, New York: Oxford University Press

Chapter 12

Policy implications of medical tourism development in destination countries: revisiting and revising an existing framework by examining the case of Jamaica

Rory Johnston[1*], Valorie A. Crooks[1] and Meghann Ormond[2]

ABSTRACT

Background

Medical tourism is now targeted by many hospitals and governments worldwide for further growth and investment. Southeast Asia provides what is perhaps the best documented example of medical tourism development and promotion on a regional scale, but interest in the practice is growing in locations where it is not yet established. Numerous governments and private hospitals in the Caribbean have recently identified medical tourism as a priority for economic development. We explore here the projects, activities, and outlooks surrounding medical tourism and their anticipated economic and health sector policy implications in the Caribbean country of Jamaica. Specifically, we apply Pocock and Phua's previously-published conceptual framework of policy implications raised by medical tourism to explore its relevance in this new context and to identify additional considerations raised by the Jamaican context.

Methods

Employing case study methodology, we conducted six weeks of qualitative fieldwork in Jamaica between October 2012 and July 2013. Semi-structured interviews with health, tourism, and trade sector stakeholders, on-site visits to health and tourism infrastructure, and reflexive journaling were all used to collect a comprehensive dataset of how medical tourism in Jamaica is being developed. Our analytic strategy involved organizing our data within Pocock and Phua's framework to identify overlapping and divergent issues.

Results

Many of the issues identified in Pocock and Phua's policy implications framework are echoed in the planning and development of medical tourism in Jamaica. However, a number of additional implications, such as the involvement of international development agencies in facilitating interest in the sector, cyclical mobility of international health human resources, and the significance of health insurance portability in driving the growth of international hospital accreditation, arise from this new context and further enrich the original framework.

Conclusions

The framework developed by Pocock and Phua is a flexible common reference point with which to document issues raised by medical tourism in established and emerging destinations. However, the framework's design does not lend itself to explaining how the underlying health system factors it identifies work to facilitate medical tourism's development or how the specific impacts of the practice are likely to unfold.

KEYWORDS

Jamaica; Medical tourism; Health services; Health policy; Qualitative methods; International medical travel; Trade in health services; Caribbean

1. BACKGROUND

As international trade in services has become increasingly desirable, health services have been identified as a promising export sector by many national governments and business consultancies worldwide [1, 2]. Medical tourism is one form of health services export that has recently attracted considerable attention, often reported as a sector that is quickly growing and immensely valuable [3,4]. Most generally, medical tourism describes the temporary movement of a patient outside the health system of their habitual country of residence for the purpose of purchasing medical care [5]. Here we consider medical tourism to be restricted to patients who *intentionally* travel to another country primarily for medical care and pay for their care privately. We impose this restriction in order to exclude 'cross-border' patients who are referred abroad by their home health systems and thereby face a different profile of health and financial risks than do medical tourists acting on their own [6, 7, 8].

Southeast Asia provides what is perhaps the best documented regional example of medical tourism development and promotion, with numerous hospitals and national and provincial governments strategically targeting the sector for investment and further growth in the wake of the 1998 Asian Financial Crisis. Private hospitals in Singapore, Thailand, and Malaysia are all known to be attracting significant numbers of medical tourists from both within and outside the region [9, 10]. These countries' national governments have been very supportive of medical tourism, creating policies and organizations to increase the export of medical services. The policies and strategies adopted among these countries commonly include the creation of visas specifically for medical tourists [11, 12], the reduction or elimination of taxation on imported medical equipment and supplies [13, 14], incentives and/or requirements for international hospital accreditation [13], and international marketing efforts that advertise the high quality of medical care available [13, 15]. When taken together, these initiatives demonstrate a regional concentration of similar promotional and development initiatives among proximate health care markets competing for international privately-paying patients.

The Caribbean provides another example of region-wide interest in medical tourism. While Cuba has long supported its health system by treating medical tourists, its early pursuit of the sector was initiated by the collapse of economic support from the Soviet Union [13]. St. Lucia, Barbados, the Cayman Islands, and Jamaica are among the countries in the Caribbean whose governments and hospitals have more recently identified medical tourism as a strategic priority for economic development [16, 17, 18, 19]. These latter countries and their health care sectors share similarities in their common development of facilitative frameworks for private health sector investment, their formation of national committees to guide the development of policies that support medical tourism, their participation in international trade conferences promoting medical tourism, and/or their development of new facilities whose primary purpose is to treat medical tourists [16, 17, 19]. These activities are an indication of the sustained and widespread interest across many Caribbean countries in developing their health services sectors for export. However, little is known about the perspectives, information, and goals driving these planning initiatives in the Caribbean or how a medical tourism sector is understood to interact with the existing health systems of these countries. This article provides a case study of the planning and development activities for the medical tourism sector in Jamaica using primary qualitative data collected in on-site fieldwork. This account provides insights into how and why the sector is emerging there and the implications its development for health policies.

1.1 Jamaican context

In the past five years, Jamaica has undertaken all of the medical tourism development initiatives common to Caribbean nations described above [20, 21, 22, 23, 24]. A small-island state with a population of 2.7 million, Jamaica is classified as an upper-middle-income country, although the country is deeply indebted and its economy is characterized by long-term stagnancy [25, 26]. Recreational tourism is one of the country's largest economic sectors, representing roughly 8 % of direct contribution to the country's gross domestic product in 2013 (26 % of the total including indirect and induced effects) and 7 % of direct employment [27, 28]. The tourism industry – highly enclavic, dependent on imports, and with significant levels of foreign investment – is concentrated on the western and northern coasts, with Kingston, the capital and country's economic centre, located on the south-central coast [29]. Medical tourism has been framed by local proponents as a viable and lucrative sector that should be developed as a natural extension of the country's existing tourism product [30, 31, 32].

In spite of the country's economic difficulties, Jamaica has a relatively robust, if severely under-resourced, health system. Jamaica's training programs for both physicians and nurses are internationally respected, the country has a strong foundation of primary care delivery that has secured low rates of communicable disease and infant mortality, and the country possesses an established network of secondary and tertiary care centres in urban centres [33]. This strong provider network and Jamaica's small size have resulted in geographically accessible health care throughout the country [34]. Jamaican citizens have universal access to publicly funded primary care clinics and secondary and tertiary hospitals without user fees, although most Jamaicans with private health insurance or the means to afford out-of-pocket payments prefer to access primary and secondary care at private facilities due to negative perceptions of the public system [33, 35, 36]. This is a source of inequitable access between poorer and wealthier Jamaicans, with a greater range of care options and resources available to the portion of the population with a lower burden of poor health because of their higher socio-economic status [37]. Overall, the vast majority of inpatient services are provided in public facilities while the private sector provides roughly half of the total share of outpatient services [38].

While there are few financial barriers facing Jamaicans seeking public medical care, the country's weak economy has proved a major challenge in sustaining adequate financing for the country's medical facilities across both the public and private health sectors [33, 35]. Funding shortfalls have contributed to major staffing shortages among nurses and specialists nationwide, especially as they face relatively few barriers migrating to the more lucrative health systems of Canada, the United Kingdom, and the United States [39, 40]. These health system factors serve as important context for Jamaica's developing medical tourism industry, which is focused on the country's private health care sector. Jamaica has eight private hospitals [33], although there is little third party data available on their quality, range of services, or utilization. Reports on Jamaica's overall health sector capacity do not provide detailed breakdowns of the public/private share of the country's 4,800 hospital beds or its health human resources, nor are there reliable collations of information about the private health sector in general [41].

1.2 Study rationale and goal

In this article we seek to contribute to the ongoing scholarly conversation about the impacts of medical tourism on health systems, their workers, and their local users (*e.g.* [10, 42, 43, 44]) by documenting and critically assessing the planning and perceptions surrounding the development of medical tourism in the Caribbean country of Jamaica. We do this by applying Pocock and Phua's [45] conceptual framework of the policy implications for destination countries posed by the development of medical tourism. Pocock and Phua previously developed this framework, published in*Globalization & Health,* following a retrospective review of the literature on the well-established medical tourism sectors in Malaysia, Singapore, and Thailand. Their purpose in developing this conceptual framework was not to assess the systemic elements of 'successful' health services exporters, but instead to highlight and organize common health policy implications raised by medical tourism for destination countries. Table 1 outlines the health system impacts and their associated policy implications identified by the framework, which are structured around core health system elements into the domains of governance, financing, delivery, regulation, and health human resources.

The broad scope of the health system domains and the high-level considerations identified in Pocock and Phua's framework encourage its application beyond the original context where it was developed. We use the existing framework to organize our findings from fieldwork in Jamaica that seeks to understand how medical tourism is being developed there. As Jamaica is *prospectively*developing a medical tourism sector located outside of the region where the conceptual framework was derived, we extend the framework and its implications outside its original geographic and temporal scope while also providing the first in-depth account of Jamaica's plans for medical tourism. This approach explores the wider relevance of the framework and its implications in a new context and identifies numerous additional system impacts and policy implications from the Jamaican experience that refine or are not captured by the existing Pocock and Phua framework. These include new policy implications in every existing domain of the existing framework and the creation of an entirely new domain, that of 'consumers'. Taken together, we believe our revised framework, derived from new insights gleaned from Jamaica, provides a more comprehensive range of relevant policy implications emerging from the development of medical tourism for monitoring and policy development worldwide.

Table 1. Pocock and Phua's (2011) conceptual framework for medical tourism's policy implications

	Policy Implications
Governance	• number and content of health sector commitments in multi- and bilateral trade agreements
(Legislation and Planning)	• regional trade blocs promoting trade in health services
	• national medical tourism committees or agencies
	• creation of medical tourism travel visas
Financing	• increase in out of pocket payments
(Fundraising and Payment)	• increasing interest in internationally portable health insurance
Delivery	• growth of private health sector
(Service Provision and Infrastructure)	• foreign direct investment in health infrastructure
Regulation	• public and private sector quality control
(Protocol Creation and Enforcement)	• international accreditation of health facilities (*e.g.* Joint Commission International)
	• number of medical tourist visits facilitated by brokers
Human Health Resources	• distribution of specialists between public and private health sector
(Training and Supply of Care Personnel)	• future human resource capacity (re: training, availability,professional to population ratios)

2. METHODS

Building on our existing Caribbean fieldwork [16, 42, 46], Jamaica was identified as a highly relevant site to prospectively explore how medical tourism is being pursued by Caribbean governments and health care providers that do not have an existing reputation for health services export. Ongoing monitoring of Caribbean media and online medical tourism promotion outlets such as the Medical Tourism Association and the International Medical Travel Journal indicated Jamaica's interest in developing a medical tourism sector. This was further reinforced by the presence of Jamaican representatives at the 2011 Global Healthcare Congress where the first two authors were in attendance. These initial impressions of Jamaica's promotional activities for medical tourism made it clear that the country does not have sizeable existing inflows of international patients but that its government and care providers are working to build a medical tourism sector. This indicated that the country would serve as a suitable case to explore how medical tourism is emerging in a new location using Pocock and Phua's framework to identify the health policy implications of its development.

After receiving ethics approval from Simon Fraser University's research ethics board, we travelled to Jamaica in October, 2012, for a two week period of on-site fieldwork. The first author returned to Jamaica for an additional four weeks of fieldwork in June/July, 2013. The research trips were structured within a case study methodological framework and employed a range of qualitative methods to better understand how medical tourism is currently understood and being mobilized in Jamaica. Semi-structured interviews with health system and tourism sector stakeholders; site visits to municipalities, health facilities, and recreational tourism locations; and daily researcher group journaling structured around shared reflective questions were all used to rigorously compile a comprehensive dataset.

2.1 Data collection – semi-structured interviews

While on-site, we sought to interview key-informants that could speak to the planning and development efforts for medical tourism in Jamaica as well as the existing health and/or tourism sectors. Potential interview participants were identified based on reviews of publicly available reports and news coverage of medical tourism in Jamaica, contacting individuals and organizations with roles relevant to medical tourism development, and ongoing referrals from interview participants. We sought to maximize the diversity of key-informants in order to capture a breadth of perspectives. Seventy-six potential participants and organizations were contacted and asked to follow up by phone or e-mail if they were interested in learning more about the study or scheduling a face-to-face interview at a time and location of their convenience. A letter of information outlining the purpose and scope of the study as well as its risks and benefits was provided to participants prior to the interview, whereupon consent was obtained.

Whenever possible, interviews were digitally recorded. However, six participants declined to be recorded and requested that only written notes be taken by the interviewer(s).

In total, 18 semi-structured interviews lasting from 30-60 min were conducted with representatives of public and private hospitals, extant and planned; government ministries; organizations charged with overseeing professional and regulatory roles in the medical sector; and individual medical professionals and tourism experts. Interview questions generally inquired about current strengths and challenges within the health system, medical tourism initiatives, and anticipated impacts of a medical tourism sector, though specific questions were tailored in order to speak to each participant's expertise. Table 2 provides an overview of the participant numbers and their professional domains. Note that the total number is higher than 18 due to some cross-categorization of participants.

Table 2. Participant overview

Employment Sector	Number of participants
Governance	7
Academic and Training	4
Private Health Sector	5
Public Health Sector	3
Trade and Development	2

2.2 Data collection – site visits

We based ourselves in Kingston, the Jamaican capital, during both fieldwork periods in order to facilitate interviews and visit Jamaica's core medical infrastructure. A five-day trip to the country's recreational tourism hub on the west coast was also conducted to meet with stakeholders and observe tourist and medical facilities there. Site visits were made to five existing medical facilities identified in interviews or news reports as potential locations for medical tourism, as well as two green-field locations positioned as future locations for medical tourism facilities. Site visits involved visiting surrounding communities where the (proposed) facilities are located, touring facilities to view infrastructure and see operations first-hand, and informal conversations with workers and users when appropriate. Detailed written notes were taken by the researchers during and immediately after site visits and were transcribed and uploaded to our secure shared server for later analysis.

At the conclusion of each day during the first fieldwork period, we took one hour to write reflective notes around a set of common questions in order to capture our impressions of the site visits. Following writing, we met for an additional hour to discuss the written observations and interpretations of sites visited and debrief about the details shared in interviews during that day, recording common and divergent issues and interpretations. This daily exercise sought to ensure that a comprehensive range of observations was captured and also to improve interpretative agreement. During the second fieldwork period the first author kept an independent journal that recorded impressions and observations that was later shared among the team.

2.3 Data analysis

Following each fieldwork period, all interview transcripts and field notes were compiled in a secure shared database. The database also holds all reports and news articles that were compiled in preparation for and collected during the research trips. Using a categorization scheme adapted from Pocock and Phua's framework of policy implications emerging from medical tourism, the plans, issues, and actors referred to across the full dataset were organized into their respective domains (governance, financing, delivery, regulation, health human resources) by the first author. Plans, issues, and actors that did not conform to the existing scheme were noted and categorized separately. Following this, a synthesis of the categorized and non-conforming findings was circulated to the second and third authors to confirm their interpretation and comprehensiveness and achieve consensus. This process resulted in the most consistent of the non-conforming findings being agreed on as our sixth analytical category of 'consumers'. Our findings from this domain are outlined in the section below following the results that fall within the existing five health policy domains identified by Pocock & Phua's framework.

3. RESULTS

In this section we organize our fieldwork findings according to the health system domains outlined in Pocock and Phua's framework in order to facilitate comparison and discussion. Their domains are: governance, financing, delivery, regulation, and health human resources. Table 1 (above) provides a summary of how the framework characterized each domain and its associated policy implications. While we acknowledge the clear interrelationships between these five domains, we consider each separately in order to effectively engage with Pocock and Phua's framework. Each section in our results briefly notes where the implications from the original framework are transferable to Jamaica's experience, however we focus on the issues and impacts identified from our Jamaican fieldwork that were *not* part of Pocock and Phua's original policy implications framework. We do so given their prominence in discussions with key-informants and in order to highlight and unpack these new considerations that warrant being factored into dialogues and future studies of the health policy implications raised by medical tourism.

3.1 Governance

With the exception of creating travel visa categories specifically for medical travellers, all of the policy implications resulting from the development of medical tourism outlined by Pocock and Phua's conceptual framework are also found to be relevant to Jamaica's sector planning. Our fieldwork revealed two additional impacts relevant to the governance domain not included in their original framework: 1) the expansion of ministerial responsibilities and convergence of inter-ministerial relationships; and 2) the involvement of international development organizations in initiating and sustaining interest in medical tourism. We separately expand on these two impacts below.

Jamaica's national government has actively supported the development of a medical tourism sector in the country since at least 2007, with political leaders and government ministers from both main political parties expressing their support for specific hospital projects focusing on the international patient market and the broader idea of medical tourism as a viable strategy for economic growth. Most recently, the Ministry of Industry, Investment and Commerce, the Ministry of Tourism, and the Ministry of Health have been jointly involved in laying the groundwork for a medical tourism sector. JAMPRO, an agency of the Ministry of Industry, Investment and Commerce that is responsible for attracting foreign direct investment and promoting Jamaican goods and services for export, has been the most active participant in spearheading sector development. For example, JAMPRO has recruited international consultants to assess the viability of medical tourism and directly participated in trade shows and conferences organized by the United States-based Medical Tourism Association and the regional Caribbean Export Development Agency. It has also convened national stakeholder meetings and drafted an incentive framework to develop policy and encourage private investment in the health services sector. Interview participants made it clear that JAMPRO has played the key leadership role in developing a vision for a Jamaican medical tourism sector and organizing policy development. Participants also indicated that the Ministries of Health and Tourism and the domestic private health sector, while supportive of medical tourism, have participated in a more limited consultative role rather than overseeing policy development or coordinating private investment in health facilities.

Another group of actors involved in medical tourism sector development and related to health sector governance that were raised in the interviews are international development organizations – a group not considered in Pocock and Phua's framework. In conversations with participants, the Canadian International Development Agency (CIDA) and the Commonwealth Secretariat were both identified as having played some role in advancing the medical tourism sector. These economic development agencies have been directly implicated in the development of medical tourism by helping to fund regional Caribbean conferences on health tourism more broadly and in commissioning and funding the production of strategic reports that explore how to develop medical tourism in Jamaica. A participant further noted that funding provided by such agencies earmarked for economic development might be used in the future to commission additional studies or to offset tax incentives for growth in the medical tourism sector, demonstrating the potential scope and policy relevance of this consideration.

3.2 Financing

Our fieldwork found limited consideration of medical tourism's likely influences on patterns of care financing. If discussed at all, out-of-pocket payments were understood to be the likely source of health care financing for any near-term plans for medical tourism. However, in light of Jamaica's proximity to the United States and interest in attracting American patients and insurers, intersections between international hospital accreditation and international health insurance portability not identified in the original Pocock and Phua framework were regularly raised by participants from the medical sector. Progress on international health insurance portability was reported to be undermined by the lack of local medical facilities with international hospital accreditation. This was compounded by the significant financial barriers to successfully obtaining international accreditation facing medical facilities. As such, while accessing foreign insurance funds was perceived by most participants to be a long-term goal, out-of-pocket payments are understood to be the most likely financing method for international patients seeking care in the immediate future.

3.3 Delivery

Jamaica's experience in planning for medical tourism thus far broadly confirms the applicability of Pocock and Phua's anticipated policy-relevant issues in the delivery domain. Our fieldwork revealed three new implications relevant to this domain: 1) the utilization of existing, underutilized private sector supply; 2) an anticipated increase in for-profit health care delivery; and 3) the creation of 'offshore' medical facilities with deep ties to neighbouring health systems and foreign investment. These impacts are described below.

Our fieldwork found two distinct strategic models for the initial development of the Jamaican medical tourism sector being pursued simultaneously. In the first, medical tourism will primarily serve as an additional market for the existing stock of private hospitals. Jamaica has a number of private hospitals located in urban districts with nearby international airports. Participants reported that these existing private hospitals are both locally well-regarded as reputable and having under-utilized capacity. This excess capacity was identified by participants as being well suited to meeting the demands of the international patient market because of the range and quality of services available. However, it was acknowledged that there are few international patients currently accessing these facilities outside of the Jamaican diaspora. International hospital accreditation was consistently identified as the largest barrier to immediately accessing the international market. This is because accreditation processes and the costs of the renovations they typically demand pose a large and immediate financial barrier.

A second project identified by participants as being spurred by the prospect of medical tourism involves the construction of a private wing within Cornwall Regional Hospital, an existing public hospital in Montego Bay. Planning for this new wing intends for it to be closely integrated into the infrastructure and operations of the existing public hospital, but for it to be privately financed and managed. Although it was noted by one participant that the primary reason for creating the wing is to both improve the quality of care available to tourists who injure themselves while visiting Montego Bay and to bring in medical tourists, they were careful to explain that its services would be available to private paying Jamaicans as well. It was also emphasized that the additional funds generated by private services would directly cross-subsidize a new public children's hospital affiliated with Cornwall Regional Hospital, without which the new facility could not sustainably operate.

In the second model, new hospitals would be built primarily to serve (medical) tourists. This is evidenced in two recent projects. In the first project, American Global MD, a group of American physicians headed by a Jamaican-American, proposes an entirely new 75-bed private hospital explicitly framed as a medical tourism destination financed with foreign investment and built in Montego Bay, the largest city on the west coast. This facility would be staffed by American physicians, while the local population would fill nursing and ancillary positions. The second project is spearheaded by a charitable organization in the resort town of Negril, also on the west coast. The Negril International Hospital plans to serve as the hub of a larger retirement community that will target foreign and diasporic buyers. Its staffing model mirrors American Global MD's approach, with foreign physicians supplying the specialty labour and medical tourists comprising the bulk of their cases. Local patients would be served in a charitable capacity with medical tourists subsidizing their cost of care. This specific example of seasonal and year-round retirement migration supporting a medical tourism facility was favoured among numerous participants who see the realization of Jamaica's existing plans to attract foreign and diasporic retirees to the country contingent on improving the quality of local medical care. Medical tourism is seen as providing the additional revenues required to viably sustain such health system improvements.

3.4 Regulation

Our fieldwork found that the issue of international hospital accreditation previously identified by Pocock and Phua's framework was by far the most prominent regulatory issue raised in conversations with Jamaican stakeholders. Participants who had engaged with international consultants or workshops/conferences on medical tourism reported having the importance of hospital accreditation in internationally marketing hospitals emphasized to them. However, international hospital accreditation was widely perceived to be a large and costly barrier to attracting international patients, and concerns were shared that the accreditation might be clinically redundant to local standards already in place. This perception prompted some participants to raise the possibility of intra-regional coordination of health care standards and regulation as they relate to medical tourism, an issue not considered in Pocock and Phua's framework. These same participants usually raised the example of harmonized health and safety standards developed for the spa and wellness tourism sector by the Caribbean Spa and Wellness Association as a model of how intra-regional coordination of the medical tourism sector at the pan-Caribbean scale could occur. With organizational assistance from CARICOM and funding from international development organizations, the Caribbean Spa and Wellness Association organized a committee to identify best practices from existing spa standards in North America, Europe, and Asia in order to address an existing regulatory gap. The result of this extensive review process was the creation of up-to-date health and safety regulations for spa and wellness service facilities shared across Caribbean countries. This case was presented as an example of successful regional regulation within the wider health tourism sector that could be copied in developing a regional hospital accreditation standard.

3.5 Health human resources

Discussions with Jamaican stakeholders and reviews of policy and media documents revealed that all health human resources-related policy implications identified by Pocock and Phua's framework are relevant to Jamaica's plans for an expanded medical tourism sector. We identified three additional policy-relevant issues pertaining to the health human resources domain: 1) health worker training as a marketing tool; 2) the increased overall demand for health human resources; and 3) the increasing international mobility and altered circulation of health human resources.

Jamaica's health workers were repeatedly referred to by participants as the country's greatest asset in developing the country's medical tourism sector. This confidence stemmed from the long-standing international ties underpinning local training practices and standards that have produced high quality and culturally familiar medical traditions comparable to other Commonwealth countries. This confidence was also informed by the long history of Jamaican diaspora working as physicians and nurses in the health systems of Canada, the United Kingdom, and the United States. It was thought that the established familiarity of Jamaican health workers in these target medical tourist markets and their reputable and interpretable training backgrounds would serve as a powerful marketing tool for health service exports from Jamaica.

Participants were largely unconcerned by the potential for medical tourism to reduce the availability of physicians in the public health system due to increased demand. This perception was not the same across the two different models of medical tourism being discussed (*i.e.*, domestic services export versus temporarily importing foreign health workers). With regard to the first model where Jamaican health care facilities and health workers provide the services, participants consistently reported that the existing arrangement between Jamaican private and public health care provides evidence that the two streams of delivery are not in competition, as most specialists participate in care delivery across both systems and nursing wages are comparable between them. Meanwhile, in the second model, facilities plan to recruit foreign physicians to work in Jamaica on a temporary basis at offshore hospitals focusing on medical tourists. This was seen as a way for foreign physicians and their families to go on vacation, to generate additional income or, as in the case of the Negril International Hospital, to provide charitable care for locals as well. Thus, this second model was perceived as having limited potential to impact the existing Jamaican health system because of these facilities outsourcing much of their own specialist labour.

Given the extreme nursing shortage in Jamaica induced by extraordinary out-migration, participants generally gave deep consideration to the potential impacts on nursing availability in the public system resulting from additional demand for private health care, regardless of staffing formats for physicians. Nurse migration from Jamaica was chiefly presented as being motivated by a desire to earn more income abroad. Medical tourism was generally understood to be a counter-weight that would provide higher wages domestically, incentivizing nurses to remain in Jamaica and ultimately increasing the local availability of health workers. This widely shared outlook was contrasted by a minority who cautioned that the higher wages that could be supported by private care provision for international patients could serve to attract the most qualified or senior care providers from the public sector. Altogether, participants demonstrated greater sensitivity to medical tourism's potential to impact the local availability of nurses than physicians.

3.6 Consumers

The domain of 'consumers' was not identified in Pocock and Phua's original framework. Our fieldwork in Jamaica, however, raised a number of critical policy-relevant questions about how perceptions of the international patient market are compelling policy action and, in turn, how a larger volume of health service exports might alter the composition of health system users in medical destinations. Participants demonstrated a consistent vision of the composition of the international patient market and, consequently, the patients that Jamaican stakeholders are working to attract. There was a widespread perception that Jamaica will tap into a large market of under- or un-insured patients from the United States. While Canada and the United Kingdom were sometimes mentioned as additional markets, the United States is the predominant target market shaping stakeholders' current expectations and planning efforts.

Participants typically did not identify patients from within the Latin American and Caribbean region or the Jamaican diaspora as potential medical tourists informing current planning. Once prompted, participants generally expressed enthusiasm for attracting patients from wherever they might come, but it was clear that medical tourists are commonly interpreted as being patients from high-income settings who are not emigrants with ties to the country. This clarifies who are understood to be medical tourists among sector stakeholders in Jamaica and therefore which international patients are being most actively sought and planned for.

4. DISCUSSION

The policy implications raised by the planning for medical tourism in Jamaica and their significance are discussed below within the domains of Pocock and Phua's existing framework. Table 3 provides a synthesis of our additional key considerations.

4.1 Governance

In a departure from Pocock and Phua's original framework, the involvement of Jamaica's Ministry of Tourism in planning for medical tourism sector development reinforces that, regardless of discursive debates around medical tourism versus medical travel (*e.g.* [47, 48]), the frame being used to develop policy for the sector in emerging destinations hinges on existing experience with recreational tourism. This approach is producing novel intersections between government ministries, evident in the involvement of both the Ministry of Health and Ministry of Tourism in the consultation and policy planning being led by Jamaica's foreign investment and export promotion corporation, JAMPRO. These intersections may result in incoherent or competing priorities informing the development of the sector. This was suggested by the regular conflation among many stakeholders that were interviewed between their enthusiasm and plans for a 'health tourism' product – including wellness retreats, complementary and alternative treatments – with the narrower, biomedical focus of 'medical tourism'. The Ministries involved and their respective roles demonstrate that medical tourism is being pursued primarily as an economic, not health system, development project in Jamaica. While the Ministry of Health's support for the sector may result in outcomes that benefit and integrate with the local health system, its role in developing and regulating an emergent medical tourism sector could also serve as a distraction from its core mandate of overseeing local health (care) concerns.

Table 3. Additional policy implications from medical tourism development

	Original Policy Implications	Additional Implications Identified from Jamaican Case
Governance (Legislation and Planning)	• number and content of health sector commitments in multi- and bilateral trade agreements • regional trade blocs promoting • trade in health services • national medical tourism committees or agencies • creation of medical tourism travel visas	• expanding/conflicting ministerial responsibilities and novel inter-ministerial relationships • involvement of international • development organizations and foreign for-profit industry organizations in developing medical tourism sectors
Financing (Fundraising and Payment)	• increase in out of pocket payments • increasing interest in internationally portable health insurance	• intersections between international hospital accreditation and international health insurance portability
Delivery (Service Provision and Infrastructure)	• growth of private health sector • foreign direct investment in health infrastructure	• utilization of existing private sector oversupply • increased for-profit healthcare delivery • cross-subsidization schemes to explicitly benefit locals • development of enclavic medical tourism facilities
Regulation (Protocol Creation and Enforcement)	• public and private sector quality control • international accreditation of health facilities (*e.g.* Joint Commission International) • number of medical tourist visits facilitated by brokers	• regional development and coordination of healthcare standards
Human Health Resources (Training and Supply of Care Personnel)	• distribution of specialists between public and private health sector • future human resource capacity (re: training, availability, professional to population ratios)	• health worker training as marketing tool • increasing international mobility and circulation of healthcare labour (including importation) • increased demand for different types of health human resources with varying supply
Consumers (Composition and Number of Patients)		• narrow conceptions of international patient market and inflated projections informing sector development • increased utilization of health services by emigrant diaspora

The role of international aid and development agencies as well as professional medical tourism industry associations observed in the Jamaican context was also not previously captured in Pocock and Phua's framework. We think their role most coherently fits within the domain of 'governance' given the role that the various groups and institutions have played in informing and supporting sector development. Because of the potential for medical tourism to negatively impact the equitable delivery of health care [43], the support of international development organizations for the medical tourism sector raises important questions for both their accountability and overall coherence of different initiatives' goals. Any efforts by development organizations to promote medical tourism as an economic development strategy must work to ensure that health equity is not compromised. This could be through concurrent efforts to advance initiatives such as universalizing health insurance coverage and transparent healthcare cross-subsidization schemes in jurisdictions that are encouraged to develop medical tourism.

4.2 Financing

Medical tourism's anticipated impacts on health care financing in Jamaica are consistent with those highlighted by Pocock and Phua's framework, but our results indicate financing impacts are likely to be staged according to phases of sector development. While out of pocket payments are perceived to be a matter of necessity, access to an increasingly portable American and European health insurance market was described as a key long term goal. The original framework does not capture the relationship between access to international insurance markets and international hospital accreditation that was identified by Jamaican interview participants. This widely shared perception helps to further explain the growing popularity of international hospital accreditation. Instead of only being framed as a market response to individual patient demands for recognizable seals of quality, international hospital accreditation also positions hospitals as viable participants in an emerging, internationally portable insurance environment [10]. International accreditation may be critical in anticipating current and future flows of insured international patients. Should the volume of international medical travel grow significantly, it will make it increasingly important to have a clear understanding of the strengths and limitations of the various international hospital accreditation standards that are being used to underwrite the risks of international care.

4.3 Delivery

The numerous new private hospital projects being planned for in Jamaica that are reliant on foreign patients confirm the original framework's general insight that increasing interest in medical tourism is accompanied by the growth of the private health sector. Our findings further demonstrate that medical tourism is also spurring significant *qualitative* changes to health services organization and delivery. For example, medical tourism's direct role in informing the planning for a new private wing in a public hospital with the expectation that it will cross-subsidize public access to care is not captured by the broad original consideration of a growing private health sector. Similarly, in systems such as Jamaica where participants reported excess capacity in the private health care sector (a key element in narratives explaining the shift towards the international market among Asian hospitals (*e.g.* [13]), turning to international patients in order to meet existing capacity can be interpreted as more efficient utilization of existing private resources rather than overall growth of the private sector. This alternate interpretation of how the private health sector might engage in or respond to medical tourism raises questions about how under-utilized private hospital beds inform interest in medical tourism among hospitals and governments.

In their plans to focus their services on international patients, the proposed 'offshore' hospitals in Jamaica invert the existing model of medical tourism as a niche market for established hospitals. These plans are echoed in other Caribbean projects such as Health City in the Cayman Islands and American World Clinics in Barbados. These projects raise important health equity concerns for the health systems that host them [16, 17, 18, 19]. In primarily employing foreign specialists and treating foreign patients, these facilities would present an especially severe form of two-tiered care, particularly if associated with the enclavic communities being planned for foreign retirees. The focus on increasing foreign investment in the health sector that these offshore hospitals represent is likely to be amplified by Jamaica's weak economic position. Interview participants from the existing private health sector (dominated by a mixture of not-for-profit hospitals and small, for-profit, physician-owned speciality clinics) clearly articulated concerns that they would face difficulty accessing the financial resources necessary to upgrade their facilities to the quality capable of attracting international patients, suggesting immediate future development may be driven by foreign investment.

This above contrast in local versus foreign financial capacity raises further questions regarding the scale and structure of foreign investment in health services relative to domestic ownership, not just an anticipated absolute increase. Given Jamaica's already crowded private hospital market, how will new, foreign-owned private hospitals impact existing locally-owned ones, especially with their attendant for-profit structure? Medical tourism development strategies that focus on inviting new, foreign investment could ultimately diminish domestic ownership in the private healthcare market over time and consolidate foreign ownership of health services. Contextual dynamics such as these demonstrate that while 'increased foreign direct investment in health services' is a fruitful point of departure for considering the impacts of medical tourism on a healthcare environment, providing even a rough outline of the existing private healthcare landscape prior to the development of medical tourism is critical to unpacking how such an outcome might actually unfold on the ground.

4.4 Regulation

As highlighted by Pocock and Phua's model, international hospital accreditation is closely tied to Jamaica's plans to develop a medical tourism sector. Stakeholder interviews demonstrated that established medical tourism destinations' experiences with international accreditation are informing Jamaica's policy development for the sector. International consultants and organizations promoting medical tourism have driven home the importance of obtaining international hospital accreditation to Jamaican medical tourism stakeholders by raising successful examples from Malaysia, Thailand, and Costa Rica. This suggests that international hospital accreditation's growth may have, in part, achieved self-reinforcing momentum as it becomes a core ingredient in the standard development formula being disseminated by international consultants promoting medical tourism. This requires investigation to determine if its high costs, relative to the economic conditions of the destinations pursuing them, are justified by a clinically relevant improvement in service quality. The Caribbean Spa and Wellness Association's success in creating their own set of therapeutic standards demonstrates that cooperation in regulating emerging economic sectors related to health and wellness most broadly is a viable route for Caribbean countries. Should medical tourism development follow suit and serve to initiate the development of a regional hospital accreditation scheme, the sector could ultimately standardize and improve the quality of hospital care in emerging destinations without incurring the costs of more expensive international accreditation regimes. This potential is unlikely to be realized, however, should marketing considerations and international insurance portability provide the dominant reasons for seeking accreditation.

4.5 Human resources

The original Pocock and Phua framework focuses on medical tourism's influence on specialist behaviour, particularly the potential for a larger private patient market to further incentivize specialists to reduce their participation in the public healthcare sector. This impact was largely interpreted as a non-issue by the majority of interview participants, being explained in three interrelated ways. Firstly, the largest medical tourism projects being pursued in Jamaica plan to rely on foreign specialist labour, insulating the local health system from losing its own supply of workers. Secondly, the kinds of procedures envisioned as constituting the bulk of medical tourist cases were elective, particularly plastic surgery and orthopaedic care, and thereby thought to insulate critical care specialities from negative impacts. Lastly, there was reported to be a great degree of existing participation across the public and private systems among specialists. It was reported that specialists seek out cases (and thereby payment) wherever available and rarely opt to solely practise in the private sector.

While the paucity of information on public versus private sector participation by medical specialists makes it impossible to independently confirm these perceptions above, they illustrate the outlooks driving the acceptability of medical tourism in emerging destinations. If medical tourism is understood to primarily deal in elective treatments, the kinds of specialist labour in increased private demand are not perceived as a threat to health equity. Arguably, this dominant outlook understates potential health equity impacts by neglecting to engage with the inevitable linkages between facilities, specialties, and overall health resources that occur in real-world health systems. Participants' heightened sensitivity to the potential for medical tourism to place stress on the insufficient local supply of nurses demonstrates the relevance of these linkages. Their concerns also highlight the need to differentiate the impacts of medical tourism on different health professions. Whereas Pocock and Phua's original framework focuses on medical tourism's impacts on specialists and largely ignores other groups of health workers, future research examining medical tourism's impacts should seek to understand how it influences nurses, specialists, and generalists as they vary both in supply and in the degree of control they have over where and how often they work.

Lastly, we propose the addition of 'health worker training as a marketing tool' to the 'human health resources' domain because of both how often participants emphasized the quality and familiarity of Jamaican health workers as a justification for developing medical tourism sector and given the discussion of training elsewhere in the literature as an asset to sector development (*e.g.*, [49, 50]). This consideration is significant in understanding the factors that initiate and sustain interest in developing medical tourism and a key factor driving the perceived viability of the sector in Jamaica. Furthermore, formal and informal connections between health systems, particularly international patterns of health worker training and employment, are potentially powerful predictive precursors of international patient flows.

4.6 Consumers

Our fieldwork demonstrates that the beliefs surrounding who medical tourists are and the scope of their numbers are critical in understanding the enthusiasm among countries seeking to become medical tourism destinations. While there is no doubt that there are American patients willing to travel for medical care, the widespread uptake among health and tourism sector stakeholders in Jamaica of inflated projections that have been disseminated by industry promoters and consultants such as the Medical Tourism Association and Deloitte's 2008 report on medical tourism [51] is alarming. These sources' wide-ranging estimates and implicit expectation that the American health system will continue to fail in adequately insuring its own population have been critiqued as overly narrow and inaccurate elsewhere (*e.g.* [3, 52]). Regardless, articulating who sector stakeholders are targeting when developing medical tourism is important in unpacking the likely economic returns and health system impacts that the sector will produce. For example, recent work from Lunt *et al.*[53] and other scholars examining medical tourism (*e.g.* [52, 54]) suggest that the international patient market is most accurately understood as relatively small and reliant on existing social ties between particular 'source' and 'destination' countries, rather than a large, unattached market of atomised patient-consumers with a willingness to travel anywhere for the care they want. While less economically tantalizing, broadening the popular conception of medical tourists beyond the image of the uninsured American patient or wait-listed Canadian or Briton to focus on regional and diasporic populations could promote the development of health services exports that are likely to have greater uptake while also being more relevant and accessible to the domestic population.

5. STUDY LIMITATIONS

As researchers with no personal or cultural ties to Jamaica, the issues that were raised in interviews and the manner in which they were framed by participants may be different than those produced by a local researcher. Controversial issues or factors that might negatively impact foreign perception of Jamaica (*e.g.*, high crime and violence rates) may have been omitted by participants, even unintentionally. As our findings emerged from a qualitative case study design, the policy implications raised here are not generalizable to all jurisdictions developing medical tourism sectors. This is not a true limitation, though, as qualitative research seeks to achieve transferability and not generalizability. Further to this, the considerations identified here and in the original framework are intended to raise implications for further investigation rather than serve as a prescriptive account.

6. CONCLUSIONS

This analysis demonstrates that the domains of Pocock and Phua's framework for identifying policy-relevant implications of the medical tourism sector provide a useful common reference point for exploring the development of medical tourism in new contexts. The original framework's inclusion of very specific issues (*e.g.* 'creation of medical tourism visas') alongside broad ones (*e.g.* 'growth of private health sector') and the absence of distinction between the drivers and impacts of medical tourism provides great latitude in adapting it to different locations and stages of development, including Caribbean countries like Jamaica. While its basic design does not supply explanatory power or trace the relationships between the pre-existing conditions that facilitate medical tourism and the impacts of the practice, the Pocock and Phua framework does provide a useful organizational scheme to identify and compile additional policy-relevant observations. For example, most of the additional policy implications that we identified in the Jamaican context are logical extensions of the original five domains in the framework. Our addition of 'consumers' to the framework as a new policy-relevant domain expands Pocock and Phua's original conceptualization to capture issues emerging from our work in the Jamaican context. It is likely that the numerous previously unidentified policy implications and new domain reported in the current analysis hold true for the wider Caribbean and are also relevant beyond the region.

We believe that future health services research examining medical tourism can further and meaningfully contribute to refining the Pocock and Phua framework by identifying additional implications emerging from the development and operation of the sector in other countries and regions beyond those already considered. Such research will assist in assessing the global transferability of the framework. Future studies can also advance our understanding of the policy implications of medical tourism development by undertaking comparative analyses of the implications specific to particular framework domains, which will help us to understand the role that local context plays in shaping how they are occurring in particular ways in specific places. At the same time, such research will further our understanding of common trends that are structuring the development of medical tourism worldwide and the scope of its health system impacts and policy implications despite such differences.

ACKNOWLEDGEMENTS

We are very grateful to the interview participants for their generous contribution of time and knowledge. This study was funded by a Planning Grant awarded by the Canadian Institutes for Health Research (FRN 122685). RJ holds a Banting and Best Doctoral Award from the Canadian Institutes for Health Research. VAC is funded by a Scholar Award from the Michael Smith Foundation for Health Research and holds the Canada Research Chair in Health Service Geographies.

AUTHORS' CONTRIBUTIONS

RJ and VAC conceived of original study, and all authors participated in study design and implementation. RJ, VAC, and MO participated in identifying potential participants, conducting fieldwork, data collection and analysis. RJ drafted the original manuscript, with VAC and MO contributing edits in following iterations of writing. All authors have reviewed and approved this manuscript.

REFERENCES

1. Lautier M. Export of health services from developing countries: the case of Tunisia. Soc Sci Med. 2008;67:101–10.
2. Smith RD, Chanda R, Tangcharoensathien V. Trade in health-related services. Lancet. 2009;373:593–601.
3. Connell J. Tourism Manage. 2013;34:1–13.
4. Horowitz MD, Rosensweig JA, Jones CA. Medical tourism: globalization of the healthcare marketplace. Med Gen Med. 2007;9(4):33.
5. Ormond M. Medical tourism. In The Wiley-Blackwell Companion to Tourism. 2nd edition. Edited by Lew AA, Hall CM, Williams A. Malden, MA: John Wiley & Sons; 2014:425-434.
6. Glinos IA, Baeten R, Maarse H. Purchasing health services abroad: Practices of cross border contracting and patient mobility in six European countries'. Health Policy. 2012;95(2):103–12.
7. Lunt N, Carrera P. Medical tourism: assessing the evidence on treatment abroad. Maturitas. 2010;66:27–32.
8. Penney K, Snyder J, Crooks VA, Johnston R. Risk communication and informed consent in the medical tourism industry: A thematic content analysis of Canadian broker websites. BMC Med Ethics. 2011;12:17.
9. Heung VCS, Kucukusta D, Song H. A conceptual model of medical tourism: implications for future research. J Travel Tourism Marketing. 2010;27:236–51.
10. Whittaker A. Pleasure and pain: medical travel in Asia. Global Pub Health. 2008;3:271–90.

11. International Medical Travel Journal. New visa exemptions to attract more medical tourists to Thailand. Int Med Travel J 2013. [http://www.imtj.com/news/?entryid82=413889]

12. Yuen M. Malaysia's visa change is a boost for healthcare tourism. eTurbo News (2009) at <http://www.eturbonews.com/13394/malaysias-visa-change-boost-healthcare-tourism>

13. Chee HL. Medical tourism and the state in Malaysia and Singapore. Global Soc Policy. 2010;10:336–57.

14. Turner L. 'First world health care at third world prices': globalization, bioethics and medical tourism. Bio Societies. 2007;2:303–25.

15. Ormond M. Neoliberal Governance and International Medical Travel in Malaysia. Abingdon: Routledge; 2013.

16. Adams K, Snyder J, Crooks V, Hoffman L. Medical tourism in the Caribbean: a call for cooperation. WIMJ Open. 2014.

17. Connell J. Medical tourism in the Caribbean islands: a cure for economies in crisis? Island Stud J. 2013;8:115–30.

18. Feinsilver JM. Fifty years of Cuba's medical diplomacy: From idealism to pragmatism. Cuban Studies. 2010;41:85–104.

19. de Arellano AB R. Medical tourism in the Caribbean. Signs. 2011; 36:289–96.

20. Jamaica Observer. Boost for Ja's medical tourism plan. Jamaica Observer. September 25, 2010. [http://www.jamaicaobserver.com/news/Boost-for-Jas-medical-tourism-plan_7994302]

21. Benzler D. The 4th Annual WMT&GHC Announces Jamaica Promotions Corporation (JAMPRO) as an Exhibitor. Med Tourism Mag | Blog 2011. [http://www.medicaltourismmag.com/blog/2011/10/the-4th-annual-wmtghc-announces-jamaica-promotions-corporation-jampro-as-an-exhibitor/]

22. International Medical Travel Journal: JAMAICA: New report on medical tourism in Jamaica. Int Med Travel J 2011. [http://www.imtj.com/news/?EntryId82=291763]

23. Braham A. PM Encourages Investment in Medical Tourism. Jamaica Information Service 2014. [http://jis.gov.jm/pm-encourages-investment-medical-tourism/]

24. Jamaica Promotions Corporation. Annual Report 2012/2013. Kingston Jamaica Promotions Corporation; 2013. [http://www.jamaicatradeandinvest.org/sites/default/files/annual_reports/JAMPRO-AnnualReport2012_2013.pdf]

25. Tennant D. Factors impacting on whether and how businesses respond to early warning signs of financial and economic turmoil: Jamaican firms in the global crisis. J Econ Business. 2011;63:472–91.

26. World Bank: Jamaica. Jamaica Data 2013. [http://data.worldbank.org/country/Jamaica]

27. World Travel and Tourism Council. Travel & Tourism Economic Impact 2014: Jamaica. London; 2014. [http://www.wttc.org/~/media/files/reports/economic%20impact%20research/country%20reports/jamaica2014.ashx]

28. Holzner M. Tourism and economic development: The beach disease? Tour Manage. 2008;32(4):922–33.

29. Williams DA, Deslandes D. Motivation for service sector foreign direct investments in emerging economies: insights from the tourism industry in Jamaica. Round Table: Commonwealth J Int Affairs. 2008;97(396):419–37.

30. Brown I. Jamaica going for a piece of medical tourism market. Jamaica Observer 2011. [http://www.jamaicaobserver.com/news/Jamaica-going-for-piece-of-medical-tourism-market_8821139]

31. Cunningham A. Jamaica Must Focus More on Health Tourism - Mascoll. Jamaica Gleaner 2013. [http://jamaica-gleaner.com/gleaner/20130617/lead/lead2.html]

32. Jamaica Information Service: Jamaica to Benefit From New Venture in Health Tourism. Jamaica Information Service 2007. [http://jis.gov.jm/jamaica-to-benefit-from-new-venture-in-health-tourism/]

33. Plummer JM, Roberts PO, Leake PA, Mitchell DI. Surgical care in Jamaica in the Laparoendoscopic Era: challenges and future prospects for Developing Nations. Perm J. 2011;15:57–61.

34. Pan American Health Organization. Jamaica. Regional Outlook and Country Profiles. Washington, DC; 2012 [http://www.paho.org/hq/index.php?option=com_docman&task=doc_view&gid=25344&Itemid=]

35. Scott E, Theodore K. Measuring and explaining health and health care inequalities in Jamaica, 2004 and 2007. Rev Panam Salud Públ. 2013;33:116–21.

36. Bourne PA. Self-reported health and medical care-seeking behaviour of uninsured Jamaicans. N Am J Med Sci. 2010;2(2):71–80.

37. Scott E, Theodore K. Measuring and explaining health and health care inequalities in Jamaica, 2004 and 2007. Rev Panam Salud Publica. 2013;33(2):116–21.

38. Anderson-Jackson L, McGrowder DA, Bourne PA, Crawford T, Whittaker WHA. Response of patients to the introduction of a private magnetic resonance imaging service in Western Jamaica. N Am J Med Sci. 2009;1(5):279–84.

39. Cawich S et al. Leadership in surgery for public sector hospitals in Jamaica: strategies for the operating room. Perm J. 2013;17:e121–5.

40. Salmon ME, Yan J, Hewitt H, Guisinger V. Managed migration: the Caribbean approach to addressing nursing services capacity. Health Serv Res. 2007;42(3):1354–72.

41. Tomblin Murphy G, MacKenzie A, Guy-Walker J, Walker C. Needs-based human resources for health planning in Jamaica: using simulation modelling to inform policy options for pharmacists in the public sector. Hum Resour Health. 2014;12:67.

42. Johnston R, Crooks VA. Medical tourism in the Caribbean region: a call to consider environmental health equity. W Indian Med J. 2013;62:250–3.

43. Hopkins L, Labonté R, Runnels V, Packer C. Medical tourism today: What is the state of existing knowledge? J Public Health Policy. 2010;31:185–98.

44. Lautier M. International trade of health services: Global trends and local impact. Health Policy. 2014.

45. Pocock NS, Phua KH. Medical tourism and policy implications for health systems: a conceptual framework from a comparative study of Thailand, Singapore and Malaysia. Globalization Health. 2011;7:12.

46. Johnston R, Crooks VA, Snyder J, Whitmore R. ''The major forces that need to back medical tourism were...in alignment'': 'championing development of barbados' medical tourism sector. Int J Health Serv, Serv. 2015;45(2):334–52.

47. Kangas B. Complicating common ideas about medical tourism: gender, class, and globality in yemenis' international medical travel. Signs. 2011;36:327–32.

48. Turner L. Transnational medical travel. Camb Q Healthc Ethic. 2013;22:170–80.

49. Ormond M. Harnessing "diasporic" medical mobilities. I. In: Thomas F, Gideon J, edltors. Migration, Health and Inequality. London: Zed Books; 2013. p. 150–62.

50. Walton-Roberts M. International migration of health professionals and the marketization and privatization of health education in India: from push-pull to global political economy. Soc Sci Med, in press.

51. Deloitte Center for Health Solutions. Medical Tourism: Consumers in Search of Value. 2008. Washington, DC. [https://www.academia.edu/9144718/Medical_Tourism_Consumers_in_Search_of_Value_Produced_by_the_Deloitte_Center_for_Health_Solutions]

52. Ormond M, Sulianti D. More than medical tourism: lessons from Indonesia and Malaysia on South–South intra-regional medical travel. Current Issues Tourism. 2014.

53. Lunt N, Jin KN, Horsfall DG, Hanefeld J. Insights on medical tourism: markets as networks and the role of strong ties. Korean J Soc Sci. 2014;41(1):19–37.

54. Noree T, Hanefeld J, Smith R. UK medical tourists in Thailand: they are not who you think they are. Globalization Health. 2014;10:29.

Index